Collins

INTERNATIONAL
PRIMARY
SCIENCE

Teacher's Guide 5

William Collins' dream of knowledge for all began with the publication of his first book in 1819. A self-educated mill worker, he not only enriched millions of lives, but also founded a flourishing publishing house. Today, staying true to this spirit, Collins books are packed with inspiration, innovation and practical expertise. They place you at the centre of a world of possibility and give you exactly what you need to explore it.

Collins. Freedom to teach.

Published by Collins
An imprint of HarperCollins*Publishers* Ltd.
The News Building
1 London Bridge Street
London
SE1 9GF

Browse the complete Collins catalogue at www.collins.co.uk

10 9 8 7 6 5 4 3

ISBN: 978-0-00-758624-0

Contributing authors: Daphne Paizee, Karen Morrison, Tracey Baxter, Sunetra Berry, Pat Dower, Helen Harden, Pauline Hannigan, Anita Loughrey, Emily Miller, Jonathan Miller, Anne Pilling, Pete Robinson.

The exam-style questions and sample answers used in the Assessment Sheets have been written by the author.

British Library Cataloguing in Publication Data
A Catalogue record for this publication is available from the British Library.

Commissioned by Elizabeth Catford
Project managed by Karen Williams
Design and production by Ken Vail Graphic Design

Acknowledgements
The publishers wish to thank the following for permission to reproduce photographs.
Every effort has been made to trace copyright holders and to obtain their permission for the use of copyright materials. The publishers will gladly receive any information enabling them to rectify any error or omission at the first opportunity.

COVER: Macro Brazilian grasshopper / Shutterstock.com

All other photos Shutterstock.

FSC is a non-profitinter national organisation established to promote the responsible management of the world's forests. Products carrying the FSC label are independently certifi ed to assure consumers that they come from forests that are managed to meet the social, economic and ecological needs of present and future generations, and other controlled sources.

Printed and bound in Great Britain by CPI Group (UK) Ltd, Croydon, CR0 4YY

Contents

Introduction

About *Collins International Primary Science*

Collins International Primary Science is specifically written to fully meet the requirements of the Cambridge Primary Science curriculum framework from Cambridge International Examinations and the material has been carefully developed to meet the needs of primary science students and teachers in a range of international contexts.

Content is organised according to the three main strands: Biology, Chemistry and Physics and the skills detailed under the Scientific Enquiry strand are introduced and taught in the context of those areas.

All course materials make use of the fully-integrated digital resources available on the DVD. For example, video clips and slideshows allow students the opportunity to view at first-hand examples of habitats, plants and animals they may not be familiar with from their own country. The interactive activities provide a valuable teaching resource that will engage the students and consolidate learning.

Components of the course

For each of Stages 1 to 6 as detailed in the Cambridge Primary Science Framework, we offer:

- A full colour, highly illustrated and photograph rich Student's Book
- A write-in Workbook linked to the Student's Book
- This comprehensive Teacher's Guide with clear suggestions for using the materials, including the electronic components of the course
- A DVD which contains slideshows, video clips, additional photographs and interactive activities for use in the classroom.

Approach

The course is designed with student-centred learning at its heart. The students conduct investigations with guidance and support from their teacher. Their investigations respond to questions asked by the teacher or asked by the students themselves. They are practical and activity-based, and include observing, questioning, making and testing predictions, collecting and recording simple data, observing patterns and suggesting explanations. Plenty of opportunity is provided for the students to consolidate and apply what they have learned and to relate what they are doing in science to other curriculum areas and the environment in which they live.

Much of the students' work is conducted as paired work or in small groups, in line with international best practice. Activities are designed to be engaging for students and to support teachers in their assessment of student progress and achievement. Each lesson is planned to support clear learning objectives and outcomes, to provide students and teachers with a good view of the learning. The activities within each unit provide opportunities for oral and written feedback by the teacher, peer teaching and peer assessment within small groups.

Throughout the course, there is wide variety of learning experiences on offer. The materials are structured so that they do not impose a rigid structure but rather provide a range of options linked to the learning objectives. Teachers are able to select from these to provide an interesting, exciting and appropriate learning experience that is suited to their particular classroom situations.

Differentiation

Differentiation is clearly built into the lesson plans in this Teacher's Guide and levels are indicated against the Student's Book activities. You will see that the practical activities offer three levels of differentiated demand. The square activities are appropriate for the level of nearly all of the students. The circle questions are appropriate for the level of most of the students (this is the level students should be achieving for this stage). The triangle questions are appropriate for some students of higher ability. Teachers may find that achievement levels vary for different content strands and interest levels. So students who are working at the circle level in Biology may find Chemistry or Physics topics more interesting and/or easier, so they may work at a different level for some of the time.

Teacher's Guide

Each double-page spread covers one unit in the Student's Book. Each unit has a clear structure identified by the *Introduction–Teaching and learning activities–Consolidate and review* sequence.

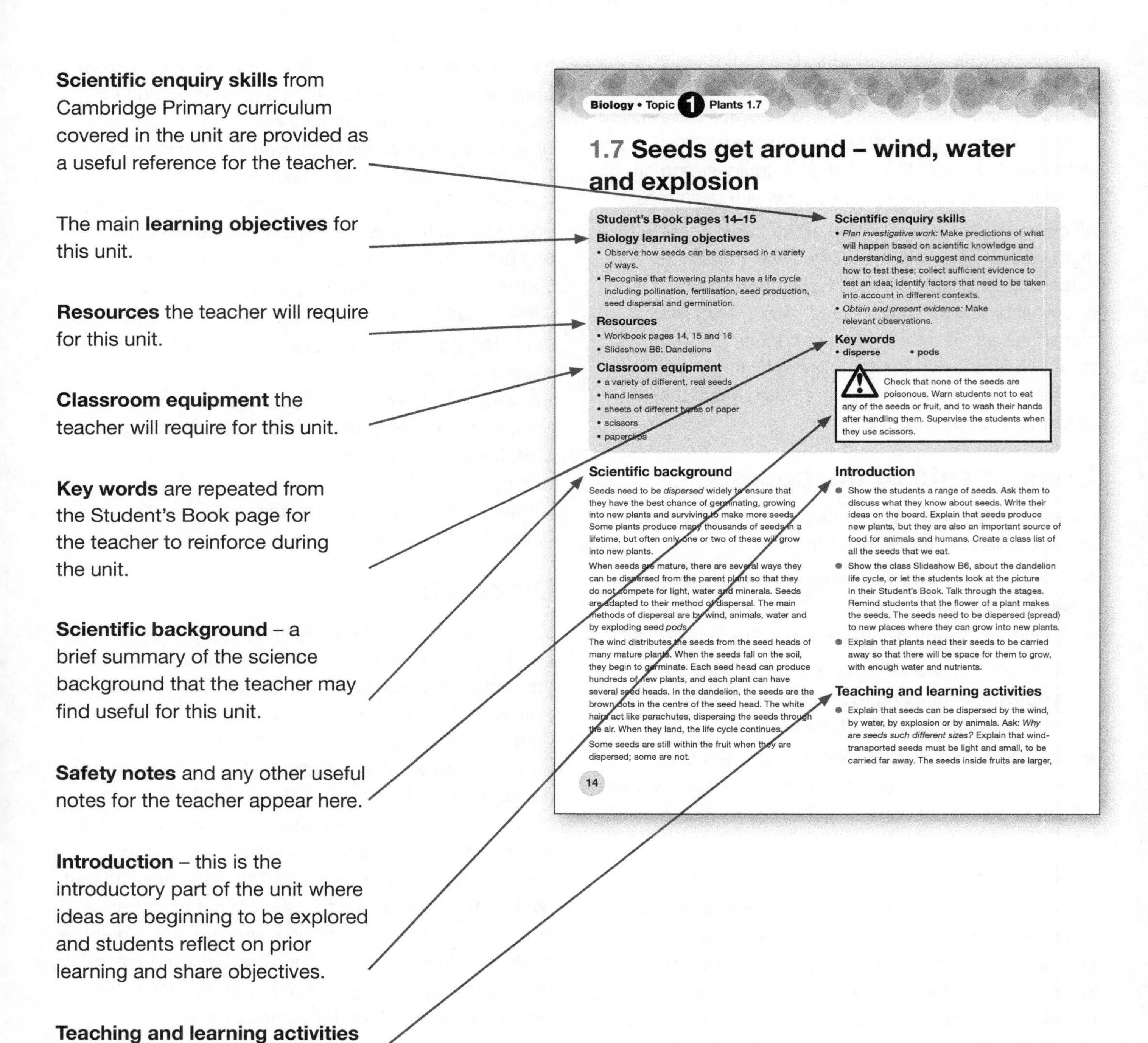

Scientific enquiry skills from Cambridge Primary curriculum covered in the unit are provided as a useful reference for the teacher.
The main **learning objectives** for this unit.
Resources the teacher will require for this unit.
Classroom equipment the teacher will require for this unit.
Key words are repeated from the Student's Book page for the teacher to reinforce during the unit.
Scientific background – a brief summary of the science background that the teacher may find useful for this unit.
Safety notes and any other useful notes for the teacher appear here.
Introduction – this is the introductory part of the unit where ideas are beginning to be explored and students reflect on prior learning and share objectives.
Teaching and learning activities – this leads into the main lesson.

Biology • Topic 1 Plants 1.7

1.7 Seeds get around – wind, water and explosion

Student's Book pages 14–15

Biology learning objectives

- Observe how seeds can be dispersed in a variety of ways.
- Recognise that flowering plants have a life cycle including pollination, fertilisation, seed production, seed dispersal and germination.

Resources

- Workbook pages 14, 15 and 16
- Slideshow B6: Dandelions

Classroom equipment

- a variety of different, real seeds
- hand lenses
- sheets of different types of paper
- scissors
- paperclips

Scientific enquiry skills

- *Plan investigative work:* Make predictions of what will happen based on scientific knowledge and understanding, and suggest and communicate how to test these; collect sufficient evidence to test an idea; identify factors that need to be taken into account in different contexts.
- *Obtain and present evidence:* Make relevant observations.

Key words

- disperse
- pods

Check that none of the seeds are poisonous. Warn students not to eat any of the seeds or fruit, and to wash their hands after handling them. Supervise the students when they use scissors.

Scientific background

Seeds need to be *dispersed* widely to ensure that they have the best chance of germinating, growing into new plants and surviving to make more seeds. Some plants produce many thousands of seeds in a lifetime, but often only one or two of these will grow into new plants.

When seeds are mature, there are several ways they can be dispersed from the parent plant so that they do not compete for light, water and minerals. Seeds are adapted to their method of dispersal. The main methods of dispersal are by wind, animals, water and by exploding seed *pods*.

The wind distributes the seeds from the seed heads of many mature plants. When the seeds fall on the soil, they begin to germinate. Each seed head can produce hundreds of new plants, and each plant can have several seed heads. In the dandelion, the seeds are the brown dots in the centre of the seed head. The white hairs act like parachutes, dispersing the seeds through the air. When they land, the life cycle continues.

Some seeds are still within the fruit when they are dispersed; some are not.

Introduction

- Show the students a range of seeds. Ask them to discuss what they know about seeds. Write their ideas on the board. Explain that seeds produce new plants, but they are also an important source of food for animals and humans. Create a class list of all the seeds that we eat.
- Show the class Slideshow B6, about the dandelion life cycle, or let the students look at the picture in their Student's Book. Talk through the stages. Remind students that the flower of a plant makes the seeds. The seeds need to be dispersed (spread) to new places where they can grow into new plants.
- Explain that plants need their seeds to be carried away so that there will be space for them to grow, with enough water and nutrients.

Teaching and learning activities

- Explain that seeds can be dispersed by the wind, by water, by explosion or by animals. Ask: *Why are seeds such different sizes?* Explain that wind-transported seeds must be light and small, to be carried far away. The seeds inside fruits are larger,

14

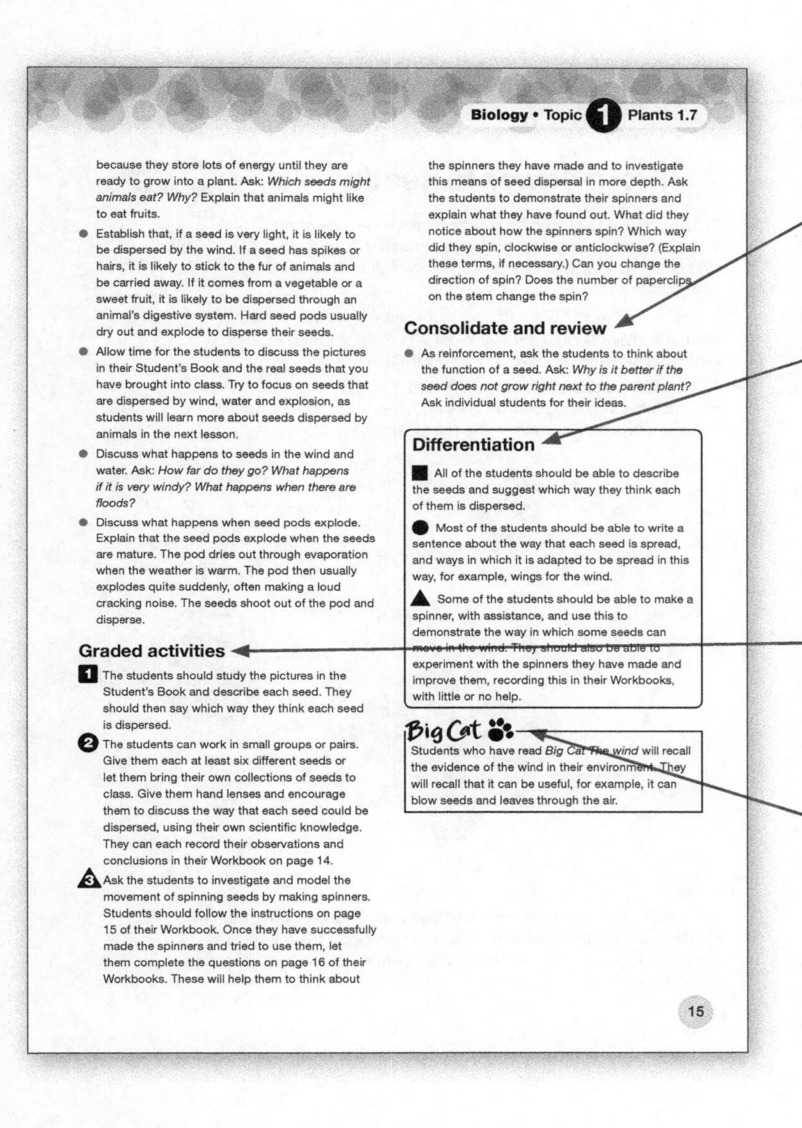

Biology • Topic 1 Plants 1.7

because they store lots of energy until they are ready to grow into a plant. Ask: *Which seeds might animals eat? Why?* Explain that animals might like to eat fruits.

- Establish that, if a seed is very light, it is likely to be dispersed by the wind. If a seed has spikes or hairs, it is likely to stick to the fur of animals and be carried away. If it comes from a vegetable or a sweet fruit, it is likely to be dispersed through an animal's digestive system. Hard seed pods usually dry out and explode to disperse their seeds.
- Allow time for the students to discuss the pictures in their Student's Book and the real seeds that you have brought into class. Try to focus on seeds that are dispersed by wind, water and explosion, as students will learn more about seeds dispersed by animals in the next lesson.
- Discuss what happens to seeds in the wind and water. Ask: *How far do they go? What happens if it is very windy? What happens when there are floods?*
- Discuss what happens when seed pods explode. Explain that the seed pods explode when the seeds are mature. The pod dries out through evaporation when the weather is warm. The pod then usually explodes quite suddenly, often making a loud cracking noise. The seeds shoot out of the pod and disperse.

Graded activities

1 The students should study the pictures in the Student's Book and describe each seed. They should then say which way they think each seed is dispersed.

2 The students can work in small groups or pairs. Give them each at least six different seeds or let them bring their own collections of seeds to class. Give them hand lenses and encourage them to discuss the way that each seed could be dispersed, using their own scientific knowledge. They can each record their observations and conclusions in their Workbook on page 14.

3 Ask the students to investigate and model the movement of spinning seeds by making spinners. Students should follow the instructions on page 15 of their Workbook. Once they have successfully made the spinners and tried to use them, let them complete the questions on page 16 of their Workbooks. These will help them to think about the spinners they have made and to investigate this means of seed dispersal in more depth. Ask the students to demonstrate their spinners and explain what they have found out. What did they notice about how the spinners spin? Which way did they spin, clockwise or anticlockwise? (Explain these terms, if necessary.) Can you change the direction of spin? Does the number of paperclips on the stem change the spin?

Consolidate and review

- As reinforcement, ask the students to think about the function of a seed. Ask: *Why is it better if the seed does not grow right next to the parent plant?* Ask individual students for their ideas.

Differentiation

■ All of the students should be able to describe the seeds and suggest which way they think each of them is dispersed.

● Most of the students should be able to write a sentence about the way that each seed is spread, and ways in which it is adapted to be spread in this way, for example, wings for the wind.

▲ Some of the students should be able to make a spinner, with assistance, and use this to demonstrate the way in which some seeds can move in the wind. They should also be able to experiment with the spinners they have made and improve them, recording this in their Workbooks, with little or no help.

Big Cat

Students who have read *Big Cat The wind* will recall the evidence of the wind in their environment. They will recall that it can be useful, for example, it can blow seeds and leaves through the air.

15

The **Consolidate and review** section is used to reinforce the students' learning during the lesson.

Differentiation – this section discusses the differentiated learning outcomes and provides the teacher with an idea of the likely behaviours of students of different ability, referencing the square, circle and triangle icons which are used across the course.

Graded activities – these are differentiated to suit three different levels of ability. They will often involve an investigation and practical element.

Links to the **Collins Big Cat** reading scheme are provided to relate science activities to the English that the students are learning.

At the end of each Topic the answers to the Student's Book questions and Assessment Sheets are given in full.

At the back of this Teacher's Guide are the Photocopy Masters (PCMs) and Assessment Sheets. These can be photocopied and handed out to the students as necessary.

The Student's Book

Each double page spread covers one unit. Each page has photographs or graphics to provide a stimulus for discussions and questions.

Key words – these are the words that the students will learn and use for this unit.

Questions – These can be used as whole class discussion points and also to enable the teacher to assess how well individual students understand the unit.

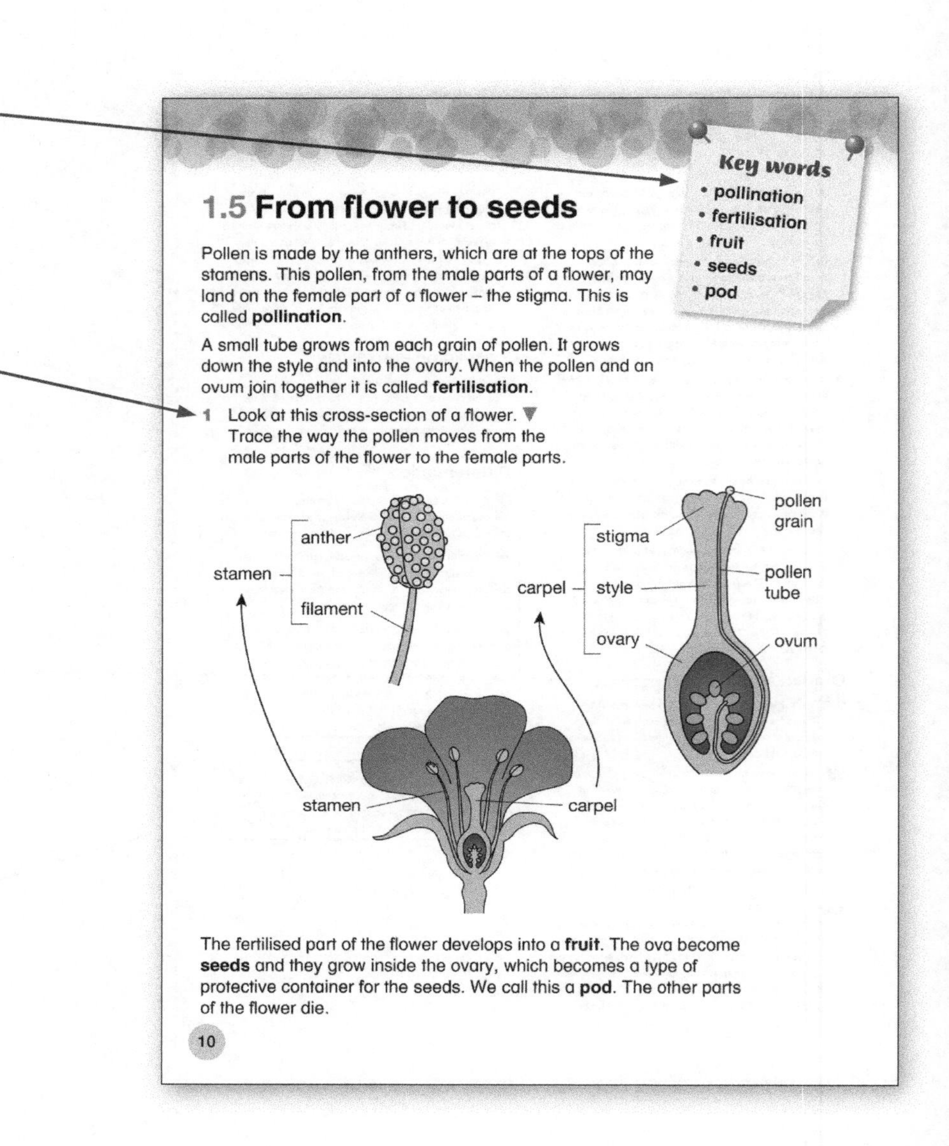

Key words
- pollination
- fertilisation
- fruit
- seeds
- pod

1.5 From flower to seeds

Pollen is made by the anthers, which are at the tops of the stamens. This pollen, from the male parts of a flower, may land on the female part of a flower – the stigma. This is called **pollination**.

A small tube grows from each grain of pollen. It grows down the style and into the ovary. When the pollen and an ovum join together it is called **fertilisation**.

1 Look at this cross-section of a flower. ▼
Trace the way the pollen moves from the male parts of the flower to the female parts.

The fertilised part of the flower develops into a **fruit**. The ova become **seeds** and they grow inside the ovary, which becomes a type of protective container for the seeds. We call this a **pod**. The other parts of the flower die.

10

Activities

Square questions are appropriate for the level of nearly all of the students.

Circle questions are appropriate for the level of most of the students (this is the level students should be achieving for this stage).

Triangle questions are appropriate for some students of higher ability.

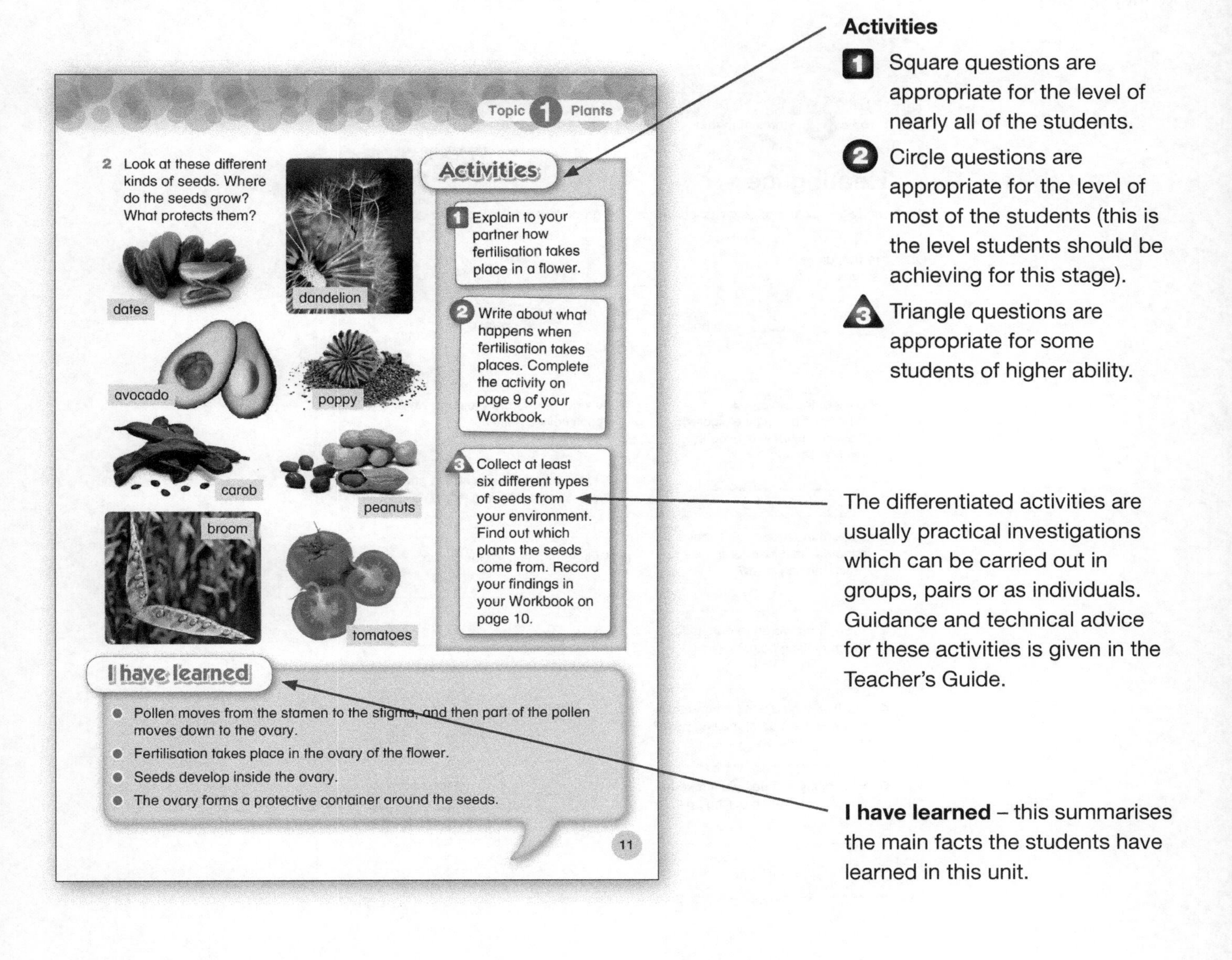

The differentiated activities are usually practical investigations which can be carried out in groups, pairs or as individuals. Guidance and technical advice for these activities is given in the Teacher's Guide.

I have learned – this summarises the main facts the students have learned in this unit.

At the back of the Student's Book is a comprehensive **Glossary** of all the Key words that are used during the lessons.

Workbook

The Workbook is for students to record observations, investigation results and key learning during the lesson. It has structured spaces for the students to record work and guidance on what to do. It gives the teacher an opportunity to give the student written feedback and becomes part of each student's work portfolio.

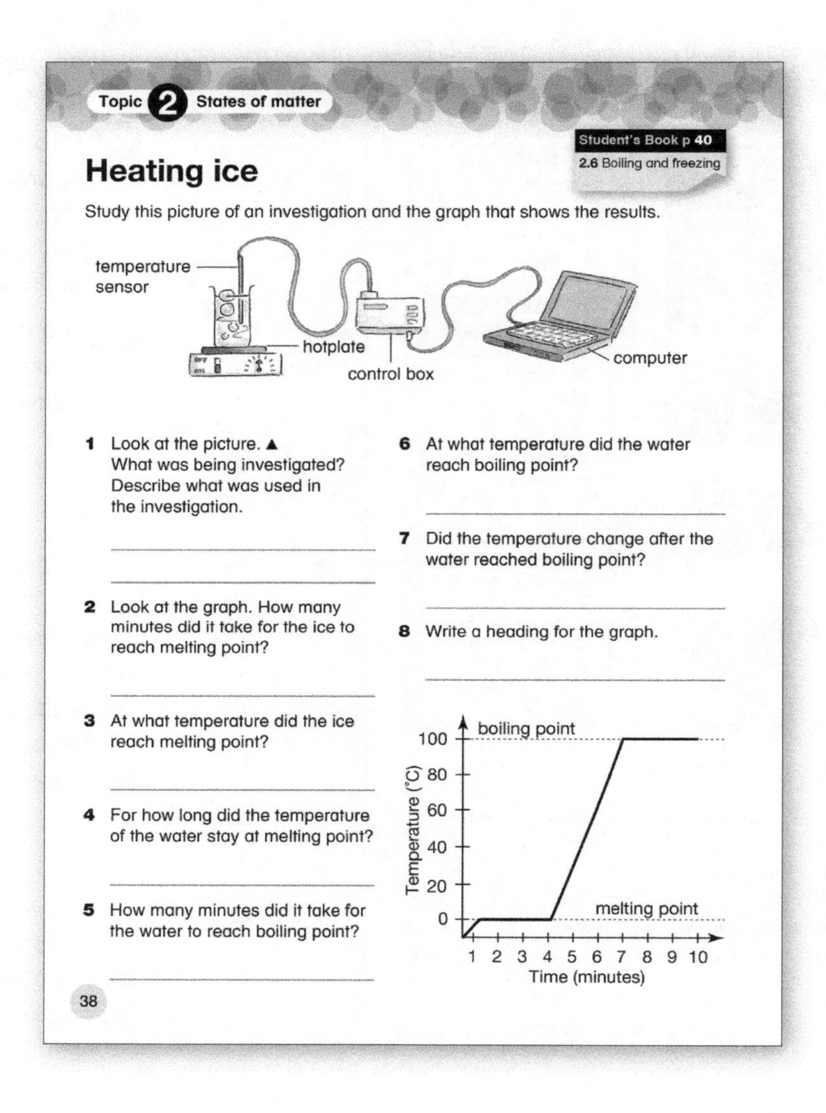

Topic 2 States of matter

Student's Book p 40
2.6 Boiling and freezing

Heating ice

Study this picture of an investigation and the graph that shows the results.

1 Look at the picture. ▲
What was being investigated?
Describe what was used in the investigation.

2 Look at the graph. How many minutes did it take for the ice to reach melting point?

3 At what temperature did the ice reach melting point?

4 For how long did the temperature of the water stay at melting point?

5 How many minutes did it take for the water to reach boiling point?

6 At what temperature did the water reach boiling point?

7 Did the temperature change after the water reached boiling point?

8 Write a heading for the graph.

38

DVD

The DVD provides teachers with a set of electronic resources to support learning and assessment. The lesson plans in this Teacher's Guide give references in the *Resources* box and in the body of text to the relevant video clips, slideshows and interactive 'drag and drop' activities.

Interactive 'drag and drop' activities

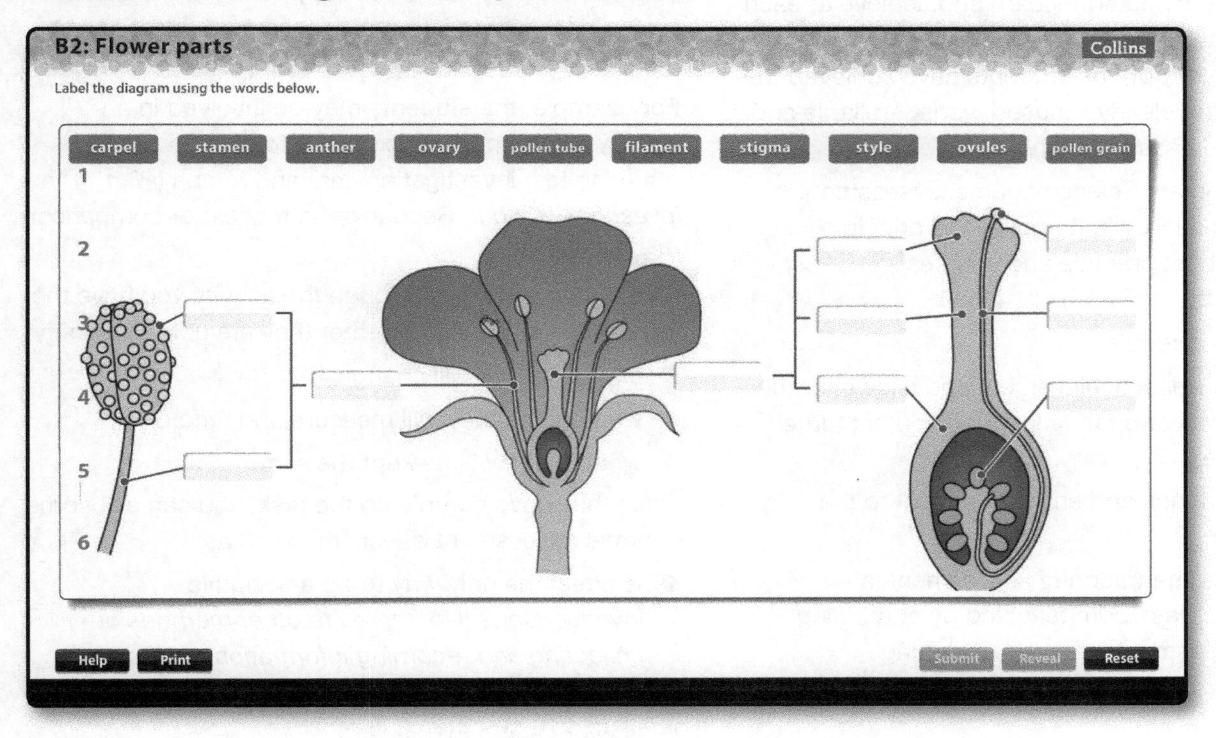

Slideshows and video clips

Assessment in primary science

In the primary science programme, assessment is a continuous, planned process that involves collecting information about student progress and learning in order to provide constructive feedback to students and parents, but also to inform planning and the next teaching steps.

Cambridge International Examinations Primary curriculum framework for science makes it clear what the students are expected to learn and achieve at each level. Our task as teachers is to make sure that we assess whether (or not) the students have achieved the stated goals using clearly-focused, varied, reliable and flexible methods of assessment.

In the Collins Primary Science course, assessment is continuous and in-built. It applies the principles of international best practice and ensures that assessment:

- is ongoing and regular
- supports individual achievement and allows for the students to reflect on their learning and set targets for themselves
- provides feedback and encouragement to the students
- allows for the integration of assessment into activities and classroom teaching by combining different assessment methods, including observations, questioning, self-assessment, formal and informal tasks
- uses strategies that cater for the variety of student needs in the classroom (language, physical, emotional and cultural), and acknowledges that the students do not all need to be assessed at the same time or in the same way
- allows for more formal summative assessment including controlled activities, tasks and class tests.

Assessing scientific enquiry skills

The development of scientific enquiry skills needs to be monitored. You need to check that the students acquire the basic skills as you teach and make sure that they are able to apply them in more complex activities and situations later on.

You can do this by identifying the assessment opportunities in different enquiry-based tasks and by asking appropriate informal assessment questions as the students work through and complete the tasks.

For example, the students may be involved in an activity where they are expected to plan and carry out a fair test investigating cars and ramps (*Plan investigative work:* Recognise that a test or comparison may be unfair).

As the students work through the activity you have the opportunity to assess whether they are able to identify:

- one thing that will change
- what things they will measure and record
- what things will be kept the same.

Once they have completed the task, you can ask some informal assessment questions, such as:

- Is a test the only way to do a scientific investigation? (*No, there are other methods of collecting and recording information, including using secondary sources.)*
- Is every test a fair test?
- Are there special things we need to do to make sure a test is fair?
- What should we do before we can carry out a fair test properly? (*Develop and write up a plan*.)
- Is a fair test in science the same as a written science or maths test at school?
- How is it different?

Formal written assessment

The Collins Primary Science course offers a selection of Assessment Sheets that teachers can use to formally assess learning and to award marks if necessary. These sheets include questions posed in different ways, questions where the students fill in answers or draw diagrams and true or false questions among others.

Below are some examples of the types of questions, provided on the Assessment Sheets. The Assessment Sheets can be found at the back of this Teacher's Guide.

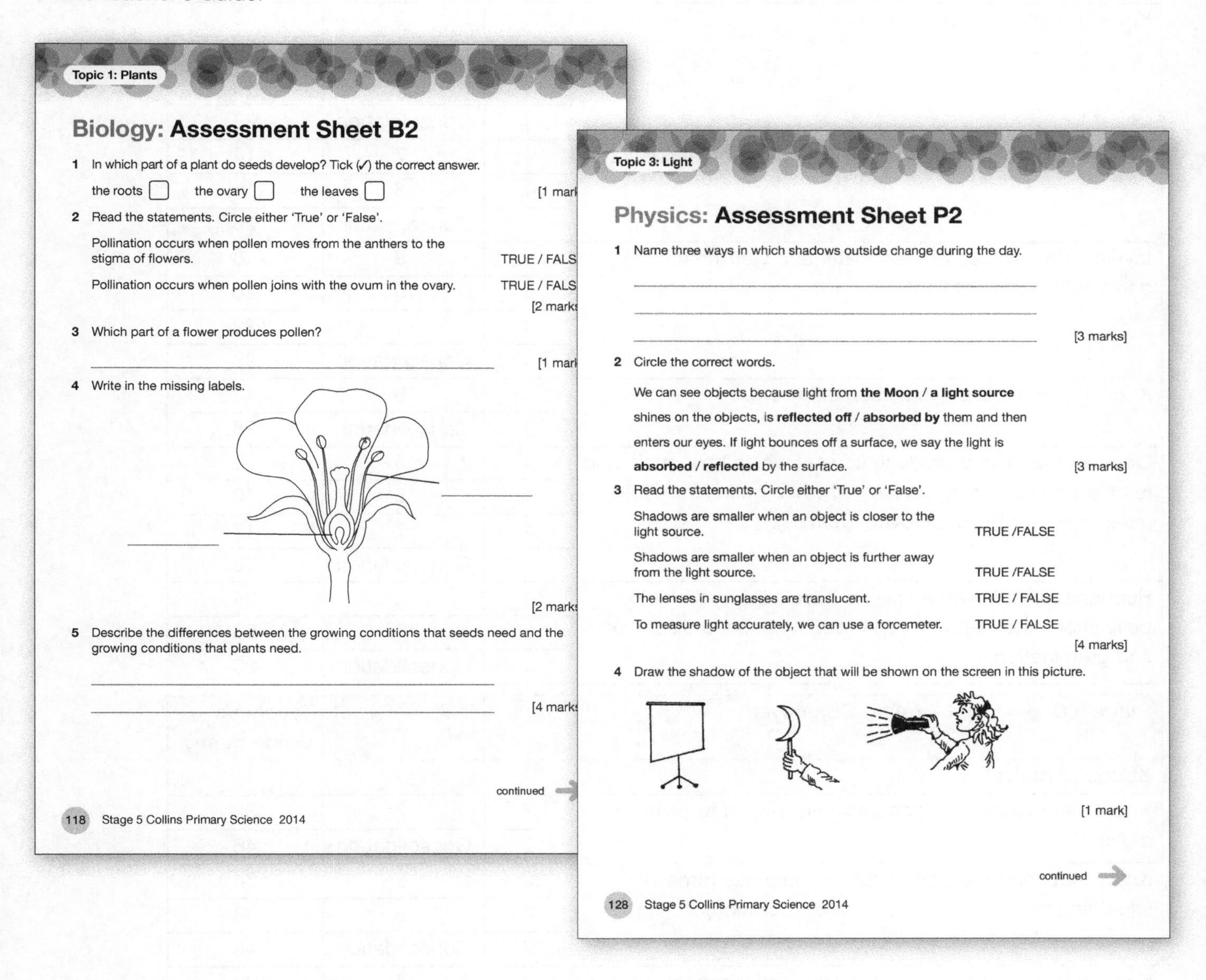

Topic 1: Plants

Biology: Assessment Sheet B2

1 In which part of a plant do seeds develop? Tick (✓) the correct answer.

the roots ☐ the ovary ☐ the leaves ☐ [1 mar

2 Read the statements. Circle either 'True' or 'False'.

Pollination occurs when pollen moves from the anthers to the stigma of flowers. TRUE / FALS

Pollination occurs when pollen joins with the ovum in the ovary. TRUE / FALS

[2 marks

3 Which part of a flower produces pollen?

______________________ [1 mar

4 Write in the missing labels.

[2 marks

5 Describe the differences between the growing conditions that seeds need and the growing conditions that plants need.

____

______________________ [4 marks

continued

118 Stage 5 Collins Primary Science 2014

Topic 3: Light

Physics: Assessment Sheet P2

1 Name three ways in which shadows outside change during the day.

____

____

______________________ [3 marks]

2 Circle the correct words.

We can see objects because light from **the Moon** / **a light source** shines on the objects, is **reflected off** / **absorbed by** them and then enters our eyes. If light bounces off a surface, we say the light is **absorbed** / **reflected** by the surface. [3 marks]

3 Read the statements. Circle either 'True' or 'False'.

Shadows are smaller when an object is closer to the light source. TRUE /FALSE

Shadows are smaller when an object is further away from the light source. TRUE /FALSE

The lenses in sunglasses are translucent. TRUE / FALSE

To measure light accurately, we can use a forcemeter. TRUE / FALSE

[4 marks]

4 Draw the shadow of the object that will be shown on the screen in this picture.

[1 mark]

continued

128 Stage 5 Collins Primary Science 2014

In addition to the materials supplied in the course, schools may opt for their students to take standardised Cambridge International Examinations progression tests at Stages 3, 4, 5 and 6. These tests are developed by Cambridge but they are written and marked in schools. Teachers download the tests and administer them in their own classrooms. Cambridge International Examinations provides a mark scheme and you can upload learners' test results and then analyse the results and create and print reports. You can also compare a learner's results against their class, school or other schools around the world and on a year-by-year basis.

Learning objectives matching grid

Stage 5 Biology Learning Objectives	Topic	Unit	Teacher's Guide pages
Plants			
Know that plants need energy from light for growth.	1	1	2
	1	2	4
	1	Consolidation	28
Know that plants reproduce.	1	3	6
	1	4	8
	1	5	10
	1	Consolidation	28
Observe how seeds can be dispersed in a variety of ways.	1	7	14
	1	8	16
	1	Consolidation	28
Investigate how seeds need water and warmth for germination, but not light.	1	9	18
	1	10	20
	1	11	22
	1	Consolidation	28
Know that insects pollinate some flowers.	1	6	12
	1	Consolidation	28
Observe that plants produce flowers which have male and female organs; seeds are formed when pollen from the male organ fertilises the ovum (female).	1	4	8
	1	5	10
	1	7	17
	1	Consolidation	28
Recognise that flowering plants have a life cycle including pollination, fertilisation, seed production, seed dispersal and germination.	1	12	24
	1	13	26
	1	Consolidation	28

Stage 5 Chemistry Learning Objectives	Topic	Unit	Teacher's Guide pages
States of matter			
Know that evaporation occurs when a liquid turns into a gas.	2	2	32
	2	Consolidation	46
Know that condensation occurs when a gas turns into a liquid and that it is the reverse of evaporation.	2	3	34
	2	5	38
	2	Consolidation	46
Know that air contains water vapour and when this meets a cold surface it may condense.	2	4	36
	2	5	38
	2	Consolidation	46
Know that the boiling point of water is 100 °C and the melting point of ice is 0 °C.	2	6	40
	2	Consolidation	46
Know that when a liquid evaporates from a solution the solid is left behind.	2	7	42
	2	8	44
	2	Consolidation	46

Stage 5 Physics Learning Objectives	Topic	Unit	Teacher's Guide pages
Light			
Observe that shadows are formed when light travelling from a source is blocked.	3	1	48
	3	10	66
	3	Consolidation	72
Investigate how the size of a shadow is affected by the position of the object.	3	3	52
	3	Consolidation	72
Observe that shadows change in length and position throughout the day.	3	2	50
	3	4	54
	3	Consolidation	72
Know that light intensity can be measured.	3	7	60
	3	8	62
	3	Consolidation	72
Explore how opaque materials do not let light through and transparent materials let a lot of light through.	3	5	56
	3	6	58
	3	Consolidation	72
Know that we see light sources because light from the source enters our eyes.	3	9	64
	3	Consolidation	72
Know that beams/rays of light can be reflected by surfaces including mirrors, and when reflected light enters our eyes we see the object.	3	10	66
	3	11	68
	3	Consolidation	72
Explore why a beam of light changes direction when it is reflected from a surface.	3	12	70
	3	Consolidation	72
The Earth and beyond			
Explore, through modelling, that the Sun does not move; its *apparent* movement is caused by the Earth spinning on its axis.	4	1	74
	4	2	76
	4	3	78
	4	Consolidation	88
Know that the Earth spins on its axis once in every 24 hours.	4	2	76
	4	3	78
	4	Consolidation	88
Know that the Earth takes a year to orbit the Sun, spinning as it goes.	4	3	78
	4	Consolidation	88
Research the lives and discoveries of scientists who explored the Solar System and stars.	4	4	80
	4	5	82
	4	6	84
	4	7	86
	4	Consolidation	88

Scientific enquiry skills matching grid

Stage 5 Scientific enquiry skills	Topic	Unit	Teacher's Guide page
Ideas and evidence			
Know that scientists have combined evidence with creative thinking to suggest new ideas and explanations for phenomena.	3	9	64
	4	1	74
	4	4	80
	4	5	82
	4	6	84
	4	7	86
Use observation and measurement to test predictions and make links.	1	1	2
	2	3	34
	2	5	38
	3	1	48
	3	2	50
	3	3	52
	3	4	54
	3	12	70
	4	2	76
	4	3	78
Plan investigative work			
Make predictions of what will happen based on scientific knowledge and understanding, and suggest and communicate how to test these.	1	7	14
	1	9	18
	1	10	20
	1	11	22
	2	1	30
	2	2	32
	2	3	34
	2	7	42
	2	8	44
	3	9	64
	4	3	78
Use knowledge and understanding to plan how to carry out a fair test.	1	2	4
	1	9	18
	1	10	20
	1	11	22
	2	2	32
	2	3	34
	2	8	44
	3	5	56
	3	6	58
	3	11	68

Stage 5 Scientific enquiry skills	Topic	Unit	Teacher's Guide page
Collect sufficient evidence to test an idea.	1	2	4
	1	7	14
	1	9	18
	1	10	20
	1	11	22
	2	3	34
	2	7	42
	3	5	56
	3	6	58
	3	11	68
	4	3	78
Identify factors that need to be taken into account in different contexts.	1	1	2
	1	4	8
	1	5	10
	1	7	14
	1	9	18
	1	10	20
	1	11	22
	1	12	24
	2	2	32
	2	3	34
	2	8	44
	3	1	48
	3	2	50
	3	3	52
	3	5	56
	3	6	58
	3	8	62
	3	11	68

Stage 5 Scientific enquiry skills	Topic	Unit	Teacher's Guide page
Obtain and present evidence			
Make relevant observations.	1	1	2
	1	2	4
	1	3	6
	1	4	8
	1	5	10
	1	6	12
	1	7	14
	1	8	16
	1	9	18
	1	10	20
	1	11	22
	1	12	24
	1	13	26
	2	1	30
	2	4	36
	2	5	38
	3	1	48
	3	2	50
	3	3	52
	3	4	54
	3	5	56
	3	6	58
	3	7	60
	3	8	62
	3	10	66
	3	11	68
	3	12	70
	4	1	74
	4	2	76
	4	3	78
Measure volume, temperature, time, length and force.	1	9	18
	1	10	20
	1	11	22
	2	6	40
	3	2	50
	3	3	52
	3	4	54
	3	6	58
	3	7	60
	3	12	70

Stage 5 Scientific enquiry skills	Topic	Unit	Teacher's Guide page
Discuss the need for repeated observations and measurements.	2	6	40
	3	2	50
	3	3	52
	3	4	54
	3	5	56
	3	6	58
	3	7	60
	3	11	68
	3	12	70
	4	3	78
Present results in bar charts and line graphs.	1	13	26
	2	2	32
	2	6	40
	3	2	50
	3	4	54
	4	3	78
	4	4	80
Consider evidence and approach			
Decide whether results support predictions.	2	2	32
	2	3	34
	2	7	42
	3	3	52
	3	5	56
	3	6	58
	3	11	68
	3	12	70
	4	3	78
Begin to evaluate repeated results.	2	6	40
	3	7	60
Recognise and make predictions from patterns in data and suggest explanations using scientific knowledge and understanding.	1	6	12
	2	6	40
	3	1	48
	3	2	50
	3	3	52
	3	7	60
	3	8	62
	3	10	66
	4	3	78
Interpret data and think about whether it is sufficient to draw conclusions.	1	8	16
	1	11	22
	2	2	32
	2	8	44

Lesson plans

1.1 Plants are living

Student's Book pages 2–3

Biology learning objective

- Know that plants need energy from light for growth.

Resources

- Workbook pages 1 and 2
- Slideshow B1: Leaves and flowers
- Video B1: Plants in the sun and rain

Classroom equipment

- small growing plant in a pot for each group of students
- rulers
- water
- pictures of a variety of different plants
- small living plants with roots, stems, leaves and flowers, in jars of water to keep them fresh

Scientific enquiry skills

- *Ideas and evidence:* Use observation and measurement to test predictions and make links.
- *Plan investigative work:* Identify factors that need to be taken into account in different contexts.
- *Obtain and present evidence:* Make relevant observations.

Key words

- **roots**
- **stems**
- **leaves**
- **flowers**
- **absorb**

If you take the students on a walk, ensure they are safe and that they stay together.

Scientific background

Plants are living things that grow. In order to grow they need soil, light and water. Plants look different but they all have *roots*, *leaves*, *stems* and *flowers*, and each part plays a specific role in the life of the plant.

In this unit the students will revise what they know about plants: the names of the different parts of plants, and what plants need to grow. Students will also revise what they know about what plants need to grow (water, soil and light) by observing a small plant growing over a period of two weeks. This will prepare them for the units that follow, in which they will learn about the process of photosynthesis and the ways in which plants reproduce.

Introduction

- Use the topic opener photograph on Student's Book page 1 to start a discussion about plants and to find out about the students' prior knowledge. Encourage the students to discuss ways in which plants are the same and ways in which they are different. Let them name the plants if they can and name any parts of the plants they can see in the picture.
- Review what students already know about plants. They should be able to name the roots, stem, flowers and leaves.
- Take the students on a tour of your school garden, if you have one. Point out the different plants and help students to name the plants. Discuss what types of leaves, flowers and fruit each plant produces. If you have a vegetable garden at school, you can also look at plants like carrots, beans, potatoes and pumpkins and identify the different parts and the parts that we eat.
- If you are unable to go outside, show Slideshow B1: Leaves and flowers. Ask the students to identify the plant parts in each slide.

Teaching and learning activities

- Show the students some of the plants you have collected or refer to the pictures in the Student's Book on page 2. Let the students examine the plants and compare them. They can describe the shapes of the roots and leaves, the colours of the stems and flowers, the shapes of the flowers and so on.
- Discuss the functions of the different parts of plants, using the questions in the Student's Book as a guide. Students should be able to work out that the roots anchor a plant in the soil. Prompt them to think about the way in which the roots *absorb* water from the soil. Then talk about where the water goes. Ask them what happens to the stems and leaves when a plant does not get water. (They droop.)

- Talk briefly about why plants have flowers. Students should have some ideas about flowers attracting insects and birds, and producing seeds. They will learn about this in more detail in later units.
- Then introduce the idea that the leaves of the plant need sunlight and water. This will prepare students for the next unit, in which they will learn about photosynthesis. Show Video B1, of plants growing in the sunshine and the rain. Ask: *What did you see in the video which shows that the plants are alive?* (They are moving and responding to the light.) *Which parts of the plants are moving? Why are the plants moving?*

Graded activities

As this is a practical investigation, it is recommended that the students work in mixed ability groups for this activity. Differentiation should be via the level of support the students receive as they work on the activity, as well as by outcome (please see guidance in the 'Differentiation' box right).

■1 The students should revise what they know about plants by completing page 1 in their Workbooks. They label the parts of a plant and describe the functions of different parts of plants.

●2 ▲3 Ask the students to work in groups to set up an investigation into what plants need to grow. Each group should have one small plant growing in a pot of soil. The students should look after the plant for a period of two weeks, watering it and making sure that it gets enough light. They record their observations, including a basic measurement of stem height, and their conclusions as to what made their plant grow (or not grow) in their Workbooks on page 2.

At the same time, students can predict what they think would happen if they placed their plant in a dark cupboard. The students will investigate whether their prediction is correct in Unit 1.2.

Consolidate and review

- Students can draw and label a different plant and write notes about the functions of different parts of a plant.

Differentiation

■ All of the students should be able to name the parts of a flowering plant and describe the basic function of each part of the plant. They should also know that plants are living things that grow, and that they need soil, water and light in order to grow. Students need to be able to recognise the parts of different plants, so make sure they see a variety of pictures and real plants.

● ▲ Most of the students should be able to observe and record their observations of growing plants. Some groups of students may need help in setting up their investigation and in recording their observations regularly. They may need support with group-work skills. Some of the students should be able to predict what will happen if a plant does not get any light.

1.2 Plants need energy from light

Student's Book pages 4–5

Biology learning objective

- Know that plants need energy from light for growth.

Resources

- Workbook pages 3 and 4
- PCM B1: Do leaves need sunlight?
- Video B2: Plants grow towards light

Classroom equipment

- two identical small bean plants of the same size for each group of students
- rulers
- a plant with several big green leaves, some thick, dark-coloured paper or cardboard, scissors, tape (for PCM B1)
- selection of leaves
- hand lenses

Scientific enquiry skills

- *Plan investigative work:* Use knowledge and understanding to plan how to carry out a fair test; collect sufficient evidence to test an idea.
- *Obtain and present evidence:* Make relevant observations.

Key words

- **photosynthesis**
- **oxygen**

⚠ Some leaves can cause irritation. Warn students not to rub their eyes or faces, and not to put their fingers near their mouths. Make sure they wash their hands after handling the leaves.

Scientific background

Like animals, plants need food to grow. Unlike animals, plants make their own food and, for this reason, animals depend on plants.

Plants need carbon dioxide and water to make their own food. Carbon dioxide is absorbed through tiny pores in the underside of the leaf. Water is absorbed through the roots and transported to the leaves through the stem.

Plants also need sunlight. A chemical called chlorophyll traps the energy from sunlight, causing a chemical reaction that produces oxygen and glucose (sugar). The glucose is transported to all of the plant's other cells, where it is used to make energy, stored as an energy source, or used to make other chemicals, such as those in nectar, pollen and oils. Students do not need to know about the process of photosynthesis as this is not part of the Cambridge Primary Science curriculum at this level and will not be tested in the Progression or Primary Checkpoint tests. The information is included here as additional information only.

Introduction

- Show students Video B2, (plants growing towards sunlight) and ask them to describe what is happening.
- Introduce the idea of the leaf being the place where the plant makes its own food. Tell the students that plants use energy from light to make their own food. Plants also need water and air to make their food. We call this process photosynthesis. (If any students ask, you can explain very briefly that during photosynthesis there is a chemical reaction in the leaves, which produces the food that the plant needs. The chlorophyll in the leaves of the plants collects the energy from the light to start this process.)
- Distribute the leaves. Ask the students to examine them and list what they can see. Give them hand lenses. Take feedback. Talk about the veins and their patterns. Explain that the veins transport water from the stem to all parts of the leaves. Talk about the pores on the underside of the leaf that let air in and out.
- Reinforce the idea that water enters the plant through the roots and travels up the stem and into the veins. Ensure that students understand that water does not enter the plant through the leaves.

- Use PCM B1 to set up a class investigation to show students what happens when the leaves of a plant do not get any light. Ask the students to predict what will happen to the covered leaf. Put the plant in a sunny spot and water it regularly. Then, after one week, let the students observe and discuss what has happened to the covered leaf.

Teaching and learning activities

- Students do not need to know or understand the process of photosynthesis at this stage, it is sufficient here to just introduce the word 'photosynthesis'. Teach the students that plants need energy from light to make their own food. If any students ask, explain in a simple way that the leaves of plants use energy from the sunlight, as well as water and air to make their food.
- Let students work in pairs and discuss what they would need to make plants grow. They should draw on prior knowledge and mention things like soil, water, light and food.
- Students look at pictures A and B of leaves on page 5 of the Student's Book and discuss the differences. They should notice the difference in colour and the fact that the leaves in picture A look healthy and the leaves in picture B do not look healthy. These differences affect a plant's ability to grow. The plant with yellow, wilting leaves will not be able to make food, so it will not be able to grow.
- Students can work in pairs. They tell their partner what they know about how a plant makes it own food. They should read the text and refer to the pictures on pages 4-5 of their Student's Books.

Graded activities

1. Students should write an article for a 'science magazine' in which they describe two ways in which animals depend on plants. They can illustrate their article with labelled pictures and use examples of animals that they are familiar with, for example goats, chickens.
2. Students should work alone to draw a picture on page 3 of their Workbooks. They should label the picture and add arrows and notes to explain what is happening in the picture.
3. Give each group of students two identical small green plants (bean plants are ideal). Ask the students to work in groups to set up an investigation to see if plants need light to grow (and, therefore, to make food). They should use the questions on page 4 of their Workbook to guide them in the investigation. They should describe what they will do to make this a fair test (by having another plant that is the same size and that is grown in all the same conditions, except sunlight), predict what the outcome will be, record their observations and measurements, and then write their conclusions.

Consolidate and review

- Students can use the pictures they have drawn in their Workbooks to explain that plants need energy from light to make food. Walk around and listen to the students. Ask questions to test understanding.

Differentiation

■ All of the students should be able to write a few sentences or paragraphs describing two ways in which animals depend on plants. Less able students may require additional help with their writing and the new vocabulary.

● Most of the students should be able to draw and label a reasonably accurate diagram to show what a plant needs to grow. They should be able to complete the activity with little or no help.

▲ Some of the students should be able to take part in setting up a fair investigation into what plants need to produce food and explain the process in more detail. They should also be able to predict what plants need to grow, and observe what happens to plants when they do not get any sunlight.

Students who have read *Big Cat Worm Looks for Lunch* should be able to identify that the animals in the story depend on plants for their food.

1.3 Plants can make new plants

Student's Book pages 6–7

Biology learning objective

- Know that plants reproduce.

Resources

- Workbook pages 5 and 6
- PCM B2: Making new plants
- Slideshow B2: No seeds

Classroom equipment

- real plant with runners, if available
- real examples of tubers, corms and bulbs
- collection of different seeds
- large sheets of paper to make posters
- coloured pens or pencils
- digital camera (optional)

Scientific enquiry skills

- *Obtain and present evidence:* Make relevant observations.

Key words

- **reproduce**
- **runner**
- **bulb**
- **corm**
- **tuber**
- **seed**

⚠ If students handle real runners/bulbs/corms/tubers/seeds, they should wash their hands afterwards. Do not allow them to taste any of these.

Scientific background

Some plants can make new plants by sending out *runners*. This is vegetative reproduction. The runners are fast growing stems that grow along the ground, rather than upwards. Roots grow down from the runners and new plantlets begin to grow at this point. Plants can spread over a large area in this way.

Bulbs are oval, or egg-shaped, and are made of layers or scales. A bulb has a pointed end that grows into a shoot. New bulbs form inside the old bulb.

Corms are rounded and flattened, with no scales. Bumps on the surface grow into shoots and new corms form around the base, as the old corm becomes exhausted.

Tubers are swollen underground roots that store food. When buried in the ground, they can grow into a new plant.

Plantlets may look like buds forming on the stem (as in cacti) or they may be small plants growing on the ends of long stems. Examples of these are shown on Slideshow B2.

Flowering plants produce fruit with *seeds*, after pollination and fertilisation has taken place. (Students will learn more about these processes in later units.)

Introduction

- Ask students how plants make new plants and let them make suggestions. They will probably suggest seeds. Then explain that some plants make new plants in another way. If possible, show the students a real plant with runners. Otherwise, ask the students to look at the pictures on page 6 of the Student's Book. Explain that a runner is a special sort of stem, which grows out from the main stem, and that a plantlet is a new plant that grows from the runner. Most plants that have runners have flowers too. Introduce or revise the word *reproduce* and remind students that all living things reproduce.

Teaching and learning activities

- Show the class Slideshow B2, about plants that make new plants without making seeds. Encourage the students to make observations and ask questions. Ensure that you describe what happens in each method of making a new plant without seeds.
- Students can work in pairs and look at any real examples of plants with runners or at the pictures of plants in their Student's Books. They should describe the plantlets and compare them to the big plants.

- Show students some of the seeds you have collected and ask them which plants the seeds come from. Ask briefly what happens to the seeds and how they grow into new plants. (You do not have to spend much time on seeds at this stage as students will learn more about them in later units.)
- Show the class any real bulbs, corms or tubers that you have collected or let them discuss the pictures in their Student's Books.

Graded activities

■1 The students should draw pictures of the real runners, seeds, and bulbs, corms or tubers in their Workbooks on page 5. Encourage them to observe the details carefully and to make their drawings scientifically accurate. They should label their drawings afterwards.

●2 Individually, the students should write sentences on page 6 of their Workbooks to describe three different ways in which plants reproduce.

▲3 The students should work in small groups or pairs to investigate reproduction in plants in their environment. Give each group or pair a large sheet of paper to make a poster. They should choose a few different examples of real plants. They will need to make accurate drawings of the plants or perhaps take photographs. They should put the plants in categories according to the way in which they reproduce and show this on the poster. Once the posters are complete, students can make presentations to the rest of the class. Let the students peer review the posters and make suggestions to improve them.

Consolidate and review

- Students work in small groups. Give each group a set of cards cut from PCM B2. Tell them first to sort the cards into four groups according to the different plants and the ways in which they reproduce. Then ask them to try and put the cards for each plant into the correct order to show how the new plants grow. Finally, they should write a short description of each of the methods of making new plants.

Differentiation

■ All of the students should be able to observe and make reasonably accurate drawings to show different forms of reproduction. Some students may be better at drawing than others, but all students should try to draw what they observe, making the plants recognisable.

● Most of the students should be able to name and describe, in a simple way, ways that different plants reproduce. They should be able to write one or two clear sentences to describe each type of reproduction.

▲ Some of the students should be able to work in groups to collect and sort information in order to make detailed, scientifically accurate drawings and descriptions of different types of reproduction. They should be able to ask and answer challenging questions and work with little assistance to produce their posters.

1.4 Flowers help plants to reproduce

Student's Book pages 8–9

Biology learning objectives

- Know that plants reproduce.
- Observe that plants produce flowers which have male and female organs; seeds are formed when pollen from the male organ fertilises the ovum (female).

Resources

- Workbook pages 7 and 8
- Slideshow B3: Flowering plant life cycle

Classroom equipment

- real flowers with both male and female parts
- cutting knife for making cross-sections
- hand lenses
- coloured pens or pencils

Scientific enquiry skills

- *Plan investigative work:* Identify factors that need to be taken into account in different contexts.
- *Obtain and present evidence:* Make relevant observations.

Key words

- **stamen**
- **anther**
- **filament**
- **pollen**
- **carpel**
- **sepal**
- **petal**
- **stigma**
- **style**
- **ovary**
- **ovum (*plural* ova)**
- **cross-section**

⚠ Students should not make cross-sections of flowers by themselves. Make the cross-sections for the students and let them watch carefully as you do it. Warn them not to do this by themselves out of school. If the students go outdoors to collect flowers, ensure they are safe and that they stay together.

Scientific background

Plants grow from seeds, which are produced by parent plants. The seeds develop inside fruits, which are formed from the flowers of the plant. The flowers grow on a plant that has grown from a seed. So, the cycle is continuous, with every plant producing seeds that will germinate into new plants. Flowers have the following parts.

Petals: these are brightly coloured. They are used to attract insects by their bright colour and scent.

Sepals: these are small green leaves around the outside of the flower. Sepals are usually smaller than the petals and are used to protect the flower while it is still in bud.

Stamens: these are where *pollen* is found. The stamens are the male parts of the flower. A stamen has two parts: the *filament* (a thin stalk) and the *anther*, which is where pollen is made.

Carpel: this is the green 'stalk' in the middle of the flower. It is the female part of the flower. The carpel has three parts: at the top is the *stigma*, which is where pollen has to land, and below the stigma are the *style* and the *ovary*, which is where seeds are formed when pollen grains join with *ova*.

Introduction

- Show the class Slideshow B3, about the life cycle of flowering plants. Students will learn more about life cycles in later units so you do not have to teach details now. Talk about the flowers and let students observe and ask questions about what they see.

Teaching and learning activities

- Show students some of the real flowers you have collected. Show the outside parts of the flowers and teach names of the parts such as petals, sepals and stamens as you work.
- Then tell students that in order to see the other parts of the flower, you are going to make *cross-sections*. Do this in front of the students. Let them come up close and examine the cross-sections. Teach the names of the parts and help students

to identify the carpel (stigma, style, ovary) and the stamens (anthers, filaments). Count the number of stamens. Give students hand lenses and let them see if they can observe any ova in the ovary. Make drawings of these cross-sections or keep them between sheets of paper to use later.

- Students can then work in pairs to identify the male and female parts of flowers as shown in the illustration in their Student's Book.
- Start a discussion about why plants need male and female parts. Students should be able to work out that the different parts have different roles.

Graded activities

1 Students work individually. They look at real flowers and at the pictures in their Student's Book, making lists of the male and the female parts of flowers. They then use these names to label the flower in their Workbook on page 7.

2 Let the students work in pairs to collect their own flowers. Help them by making cross-sections of four flowers that they have collected. Give them hand lenses and encourage them to observe the cross-sections very carefully. They should try to identify all the male and female parts. Encourage them to ask questions.

3 Make a cross-section of a flower for the students. They should then make a detailed, accurate drawing of it on page 8 of their Workbook. They should copy the size and shape of the flower carefully and label all the parts that they have observed. Let them use hand lenses to study the flower. Ask them to show their drawings to their groups or to the class and explain what they have drawn.

Consolidate and review

- Draw pictures of the cross-sections that you made at the beginning of the lesson. Write a list of the names of the different parts on the board and ask students to match the names to the different parts of the flowers. Label the pictures as you work.
- Arrange a short quiz for students or play 'Five questions' in groups. To play 'Five questions', the first player in the group thinks about a part of a flower. The others in the group can ask five questions to discover the part. The first player can only answer 'yes' or 'no' to the questions from the other players. For example: *Is it in the ovary? No. Is it a female part? No. Is it on top of the filament? Yes. Is it an anther? Yes.* The person who asks the question that provides the correct answer, continues the game by thinking about another part of a flower.

Differentiation

■ All of the students should be able to identify and list the male and female parts of a flower. They should be able to label the picture of the flower in their Workbook with little or no help. Students should be able to identify the main parts of flowers in drawings and make their own recognisable drawings, with labels.

● Most of the students should be able to identify the parts of real flowers that they have collected and that have been cut in half. They should be able to distinguish between male and female parts with some help.

▲ Some of the students should be able to observe and then make an accurate drawing of a cross-section of a flower, identifying the male and female parts and labelling their pictures accordingly, without any help.

1.5 From flower to seeds

Student's Book pages 10–11

Biology learning objective

- Know that plants reproduce.
- Observe that plants produce flowers which have male and female organs; seeds are formed when pollen from the male organ fertilises the ovum (female).

Resources

- Workbook pages 9 and 10
- Slideshow B4: Plants making seeds
- DVD Activity B1: Flower parts

Classroom equipment

- collection of different types of seeds
- cross-sections of real flowers
- real fruit with seeds (for example, tomatoes, peanuts, dates, avocados)
- coloured pens or pencils

Scientific enquiry skills

- *Plan investigative work:* Identify factors that need to be taken into account in different contexts.
- *Obtain and present evidence:* Make relevant observations.

Key words

- **fertilisation**
- **pollination**
- **seeds**
- **fruit**
- **pod**

⚠ If the students go outdoors to collect seeds, ensure they are safe and that they stay together. Students should not eat any of the seeds they collect or handle, as many seeds are treated with insecticide, which may be harmful. Wash hands after handling seeds.

Scientific background

For a seed to form, an ovum needs to be fertilised. Pollen is made in the anthers and is transferred from the anthers to the stigma. A small tube grows from each grain of pollen, down the style and into the ovary. In the ovary, the pollen fertilises the ova. *Seeds* develop from ova. A seed consists of a young plant and a reserve food supply, surrounded by a protective covering called the seed coat. Seeds develop inside a *fruit* that forms from the ovary.

Students will learn more about *pollination* and germination in later units, so focus on the process of *fertilisation* in this unit. Please note that the students do not need to know the detail of the process of fertilisation including the development of the pollen tube as this is not part of the Cambridge Primary Science curriculum and will not be tested in the Progression or Primary Checkpoint tests. It is included here for enrichment and additional information only. The students need to know that plants reproduce, and observe that plants produce flowers which have male and female organs; seeds are formed when pollen from the male organ fertilises the ovum (female).

Introduction

- Use real flowers to show the students the anthers. If possible, show them flowers with pollen on the anthers as well. Let them touch the pollen to feel that it is sticky. Ask them to tell you what they think happens to the pollen.
- Show the Slideshow B4: Plants making seeds. Let students discuss and ask questions about the seeds shown in the different plants.

Teaching and learning activities

- Use drawings or real flowers to explain to students how fertilisation occurs. You can draw a diagram of the process on the board as you explain this.
- Tell students to look at the cross-section of a flower on page 10 of their Students' Book. Explain the process of fertilisation again and let them trace the route that the pollen takes with their finger as they listen.
- Then explain what happens when the ova are fertilised and fruit develop.
- Show the students some of the real fruit you have collected. Cut or break the fruit open and let students identify the seeds and the pods. They can also look at the illustrations in their Student's Book.

- Give students time to discuss and observe the seeds. They should consider why seeds contain stored food and how the seeds are protected.

Graded activities

1 The students should work in pairs and explain to each other what happens during fertilisation. Explaining the process to another student will help them to learn the process. The students can refer to the illustrations in the Student's Book as they talk. Walk around, prompting and asking questions to check that the students understand what they are saying.

2 The students consolidate what they have learned by reading the short summary sentences about the process of fertilisation in their Workbook on page 9, and put them in the correct order. They then draw a sketch to show their own understanding of the process.

3 The students should investigate some different seeds. They should collect seeds from their local environment (if possible), draw them and compare them. They can record their observations in their Workbook on page 10. Encourage the students to examine the seeds carefully and to comment on the shape, size, texture and colour of each seed and pod. They should start to question why seeds are different and start to predict that this might have something to do with the way they will disperse and grow.

Consolidate and review

- Draw a simple flow diagram of the process of fertilisation on the board and let the students suggest how it can be completed. You could draw something like this:

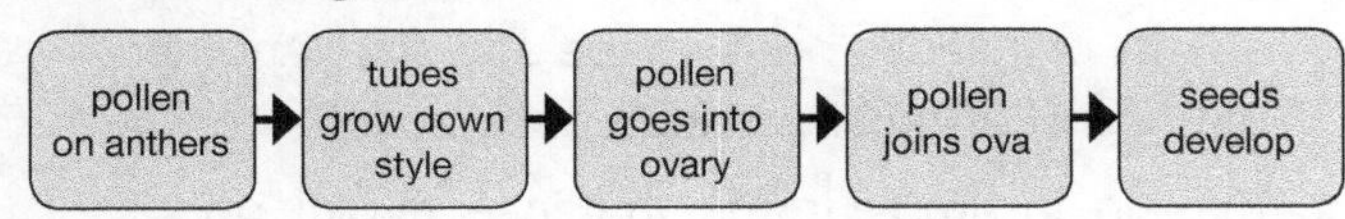

- Students can work in pairs. They can put the seeds they have collected together and compare them. They can put them in categories according to size, type of pod, or texture.
- More able students can complete DVD Activity B1 to label the different flower parts. Remind students that they do not need to know the detail of the process of fertilisation including the development of the pollen tube as this is not part of the Cambridge Primary Science curriculum and will not be tested in the Progression or Primary Checkpoint tests. It is included here for enrichment and additional information only.

Differentiation

■ All of the students should be able to identify where fertilisation takes place in a flower and explain the basic process. If students have difficulty remembering the sequence of events, write the sentences from Workbook page 9 on cards and let individual learners read the cards and put them in order.

● Most of the students should be able to put the sentences in the correct order to explain the process of fertilisation. They should be able to illustrate fertilisation with a simple sketch and explain with confidence the process by which seeds develop, from pollination to the formation of fruits.

▲ Some of the students should be able to describe and compare seeds. They should also be able to recognise that seeds develop in flowers and that seeds are not all the same. They should start to question why seeds are different and start to predict that this might have something to do with the way they will disperse and grow.

1.6 Insects and flowers

Student's Book pages 12–13

Biology learning objective

- Know that insects pollinate some flowers.

Resources

- Workbook pages 11–12 and 13
- PCM B3: Bees
- Slideshow B5: Flower adaptations
- Video B3: Bees

Classroom equipment

- range of real flowers with stamens clearly visible
- flowers of four different plants, including one that is strongly scented, one of each for each student
- coloured pens or pencils
- large sheets of paper to make posters
- access to the internet and presentation software (optional)

Scientific enquiry skills

- *Obtain and present evidence:* Make relevant observations.
- *Consider evidence and approach:* Recognise and make predictions from patterns in data and suggest explanations using scientific knowledge and understanding.

Key words

- **insects**
- **nectar**
- **pollinators**

⚠ Make sure that the students do not eat the flowers. Ensure that they wash their hands after handling the flowers. Be aware that some students may suffer from hay fever, and take necessary precautions.

Scientific background

Flowering plants depend on animals (including *insects*, birds and some mammals) as pollinators, to enable them to reproduce. Both plants and pollinators are threatened with extinction, due to habitat loss and pollution from human activities. The continuing fall in numbers of honeybees has shown the importance of insect pollinators, not only to crops consumed by humans, but also to plants that support the ecosystems on which we depend.

Introduction

- Refer back to what the students have learned about fertilisation. Ask them how they think the pollen gets from the anthers to the stigma of flowers.
- Show the class the real flowers you have collected, or let the students look at the photographs in their Student's Book. Let them observe and discuss the flowers. Make sure that they can identify the anthers and the stigma in the flowers.
- Explain that flowers need to be pollinated before fertilisation can take place. Remind students of the word 'pollination', which they met in Unit 1.5, and teach them the word *pollinator*. Remind them that in order for fertilisation to take place, pollen must first get from the anthers (the male parts) of a flower to the stigma (the female part) of a flower.

Teaching and learning activities

- Let the class watch Slideshow B5. Stress that flowers have features that indicate to insects where the *nectar* is, to entice them in and to try to ensure that, as a result, they are pollinated.
- Explain that plants need to attract insects, as they help the seeds to form. Having brightly coloured petals is one of the ways in which flowers attract insects such as bees and butterflies. The shape of the petals is often designed to suit a particular animal or insect's body shape and size. Another method of attracting insects is by scent.
- Explain, in simple terms, that insects pick up pollen on their feet and bodies when they visit a flower, and may deposit it when they settle on the next flower. This is called pollination. If the pollen fertilises this flower, it will be able to make seeds. The students may confuse pollen with seeds.
- To help students to think carefully about how flowers attract insects, let them complete the activity on pages 11–12 of their Workbook. Give each student four different flowers, including

one with a strong smell. Let them record their observations about these flowers, following the guidelines given.

- Ask students to report back to the rest of the class afterwards. Ask questions like: *Do all flowers have a smell? Are all flowers brightly coloured? Which flowers are likely to attract the most insects? Why do some plants have much bigger flowers than others? Is it better to have large flowers or smaller flowers?*
- Read the headline about bee populations in the Student's Book and let the students discuss what this means. For enrichment or homework, give the students copies of PCM B3. They can read the text and answer the questions. Take time to discuss the answers with the whole class afterwards. Show students Video B3 about bees.

Graded activities

■ 1 Students should work in pairs and explain the difference between pollination and fertilisation to their partners. Many students will find this confusing at first. Remind them that pollination is about moving the pollen from one flower to another. Fertilisation is what happens inside a flower after it has been pollinated.

● 2 Students should work in groups and discuss the importance of pollinators to gardeners and farmers. Walk around and prompt the students with questions like: *What happens when people spray flowers with pesticides? What can a gardener do to make sure that their garden is friendly to insects?*

▲ 3 Let the students do some research to produce a poster or slide presentation on the bee populations in their own environment. They may need some help with this and access to the internet if available. If possible, arrange for them to visit a local farm or garden, or invite a farmer, beekeeper or gardener to come in to talk to the students.

Consolidate and review

- Students can write their own short story or poem about insects travelling to different flowers. Ask them to give a sense of the many ways in which types of flowers differ from one another, and what features are most likely to attract the insects.
- Students can complete the activities in their Workbook on page 13 to consolidate what they have learned.

Differentiation

■ All of the students should be able to explain the difference between fertilisation and pollination with some help.

● Most of the students should be able to explain why pollinators, especially bees, are essential in gardening and farming. They will be able to say that without insect pollination many plants would be unable to grow fruits or seeds.

▲ Some of the students should be able to undertake independent research and will produce informative and interesting poster or digital presentations. They will understand that bee populations are declining and should suggest some measures being used to try and prevent this.

1.7 Seeds get around – wind, water and explosion

Student's Book pages 14–15

Biology learning objectives

- Observe how seeds can be dispersed in a variety of ways.
- Recognise that flowering plants have a life cycle including pollination, fertilisation, seed production, seed dispersal and germination.

Resources

- Workbook pages 14, 15 and 16
- Slideshow B6: Dandelions

Classroom equipment

- a variety of different, real seeds
- hand lenses
- sheets of different types of paper
- scissors
- paperclips

Scientific enquiry skills

- *Plan investigative work:* Make predictions of what will happen based on scientific knowledge and understanding, and suggest and communicate how to test these; collect sufficient evidence to test an idea; identify factors that need to be taken into account in different contexts.
- *Obtain and present evidence:* Make relevant observations.

Key words

- **disperse**
- **pods**

Check that none of the seeds are poisonous. Warn students not to eat any of the seeds or fruit, and to wash their hands after handling them. Supervise the students when they use scissors.

Scientific background

Seeds need to be *dispersed* widely to ensure that they have the best chance of germinating, growing into new plants and surviving to make more seeds. Some plants produce many thousands of seeds in a lifetime, but often only one or two of these will grow into new plants.

When seeds are mature, there are several ways they can be dispersed from the parent plant so that they do not compete for light, water and minerals. Seeds are adapted to their method of dispersal. The main methods of dispersal are by wind, animals, water and by exploding seed *pods*.

The wind distributes the seeds from the seed heads of many mature plants. When the seeds fall on the soil, they begin to germinate. Each seed head can produce hundreds of new plants, and each plant can have several seed heads. In the dandelion, the seeds are the brown dots in the centre of the seed head. The white hairs act like parachutes, dispersing the seeds through the air. When they land, the life cycle continues.

Some seeds are still within the fruit when they are dispersed; some are not.

Introduction

- Show the students a range of seeds. Ask them to discuss what they know about seeds. Write their ideas on the board. Explain that seeds produce new plants, but they are also an important source of food for animals and humans. Create a class list of all the seeds that we eat.
- Show the class Slideshow B6, about the dandelion life cycle, or let the students look at the picture in their Student's Book. Talk through the stages. Remind students that the flower of a plant makes the seeds. The seeds need to be dispersed (spread) to new places where they can grow into new plants.
- Explain that plants need their seeds to be carried away so that there will be space for them to grow, with enough water and nutrients.

Teaching and learning activities

- Explain that seeds can be dispersed by the wind, by water, by explosion or by animals. Ask: *Why are seeds such different sizes?* Explain that wind-transported seeds must be light and small, to be carried far away. The seeds inside fruits are larger, because they store lots of energy until they are ready to grow into a plant. Ask: *Which seeds might animals eat? Why?* Explain that animals might like to eat fruits.

- Establish that, if a seed is very light, it is likely to be dispersed by the wind. If a seed has spikes or hairs, it is likely to stick to the fur of animals and be carried away. If it comes from a vegetable or a sweet fruit, it is likely to be dispersed through an animal's digestive system. Hard seed pods usually dry out and explode to disperse their seeds.
- Allow time for the students to discuss the pictures in their Student's Book and the real seeds that you have brought into class. Try to focus on seeds that are dispersed by wind, water and explosion, as students will learn more about seeds dispersed by animals in the next lesson.
- Discuss what happens to seeds in the wind and water. Ask: *How far do they go? What happens if it is very windy? What happens when there are floods?*
- Discuss what happens when seed pods explode. Explain that the seed pods explode when the seeds are mature. The pod dries out through evaporation when the weather is warm. The pod then usually explodes quite suddenly, often making a loud cracking noise. The seeds shoot out of the pod and disperse.

Graded activities

■ 1 The students should study the pictures in the Student's Book and describe each seed. They should then say which way they think each seed is dispersed.

● 2 The students can work in small groups or pairs. Give them each at least six different seeds or let them bring their own collections of seeds to class. Give them hand lenses and encourage them to discuss the way that each seed could be dispersed, using their own scientific knowledge. They can each record their observations and conclusions in their Workbook on page 14.

▲ 3 Ask the students to investigate and model the movement of spinning seeds by making spinners. Students should follow the instructions on page 15 of their Workbook. Once they have successfully made the spinners and tried to use them, let them complete the questions on page 16 of their Workbooks. These will help them to think about the spinners they have made and to investigate this means of seed dispersal in more depth. Ask the students to demonstrate their spinners and explain what they have found out. What did they notice about how the spinners spin? Which way did they spin, clockwise or anticlockwise? (Explain these terms, if necessary.) Can you change the direction of spin? Does the number of paperclips on the stem change the spin?

Consolidate and review

- As reinforcement, ask the students to think about the function of a seed. Ask: *Why is it better if the seed does not grow right next to the parent plant?* Ask individual students for their ideas.

Differentiation

■ All of the students should be able to describe the seeds and suggest which way they think each of them is dispersed.

● Most of the students should be able to write a sentence about the way that each seed is spread, and ways in which it is adapted to be spread in this way, for example, wings for the wind.

▲ Some of the students should be able to make a spinner, with assistance, and use this to demonstrate the way in which some seeds can move in the wind. They should also be able to experiment with the spinners they have made and improve them, recording this in their Workbooks, with little or no help.

Big Cat

Students who have read *Big Cat The wind* will recall the evidence of the wind in their environment and know that it can be useful, for example, it can blow seeds and leaves through the air.

1.8 Seeds get around – animals

Student's Book pages 16–17

Biology learning objective

- Observe how seeds can be dispersed in a variety of ways.

Resources

- Workbook pages 17 and 18

Classroom equipment

- photographs of elephants, tree squirrels, birds and any other animals in your area that help to disperse seeds
- coloured pens or pencils

Scientific enquiry skills

- *Obtain and present evidence:* Make relevant observations.
- *Consider evidence and approach:* Interpret data and think about whether it is sufficient to draw conclusions.

Key words

- **droppings**
- **burrs**

⚠ Warn students about getting too close to any wild animals they are observing. They should not disturb the animals in any way and they should not go too close to them. Observations of animals should always be done under adult supervision.

Scientific background

Many seeds are dispersed by animals. Some seeds hook on to the fur of passing animals. These seeds have little hooks or *burrs* on them. Other seeds, especially in fruits, are eaten by animals and then passed out in the animals' *droppings*. They are not digested in the gut of the animals. Scientists have found that the digestive fluids in the stomachs of elephants actually help some seeds to germinate more quickly. As wild animals often move great distances, the seeds are dispersed widely. Insects like ants can also help to disperse seeds as they take seeds down into their nests. They eat the seed coverings and then leave the seeds to grow under the ground.

Introduction

- Show the students any pictures you have collected of animals. Try to have pictures of different types of animals: birds, large herbivores and smaller animals that burrow. Ask: *How do you think these animals help to disperse seeds? What do the animals eat? What do they do with fruits and seeds? What other animals do you know that might help to disperse seeds?*

Teaching and learning activities

- Explain that many wild animals eat fruit. For example, elephants, monkeys, bats, foxes and birds all eat fruit. The seeds inside the fruit pass through the digestive systems of the animals, but they are not digested. The seeds come out in the animals' droppings. Animals move around a lot, so the seeds move around too.
- Other animals, like tree squirrels, collect hard seeds (nuts) when they are plentiful and store them for times of the year when no seeds are available. They bury the seeds or store them in the trunks of trees, for example. This protects the seeds.
- Discuss the photographs in the Student's Book. Find out what the students know about these animals and make sure students understand how each one helps to disperse seeds.
- Have a discussion about what happens when we discard fruit pips after eating the fruit. Ask: *Should we throw pips on the ground or in a bin? Why?*
- If appropriate, you could discuss why travellers are not allowed to take plants or fruit from one country to another country. Help students to understand that the seeds may grow into new plants and that these new plants may then grow in other countries and cause damage to local environments.

Graded activities

As these activities involve a detailed practical investigation that will take place over a longer period of time, it is recommended that the students work in mixed ability groups to complete them all. Differentiation should be via the level of support the students receive as they work on the activity, as well as by outcome (please see guidance in the 'Differentiation' box right)

■ 1 The students should work in pairs and name three animals that help to disperse seeds. They should explain how each animal does this. Students can use the photographs in their Student's Book or any other material as a reference. If you have a sufficient variety of animal pictures, the students could sort them into groups to show the different ways the animals help to disperse seeds.

● 2 The students can work in groups and observe animals in their local environment. This needs to be carefully organised so that students are not placed in any danger. It can take time to observe animals, so students may need to complete this activity over a period of a few weeks. Students may observe animals near fruit trees or bushes with berries on them. Some students should already have some knowledge of the animals in their own environment and what they eat. Assist the students so that they choose suitable animals. If observation is not possible, students could do some research on animals in their environment instead, finding out what each animal eats and how each animal helps to disperse seeds. Students can record their observations in their Workbook on page 17. Encourage them to think about whether or not they have collected enough data to draw conclusions about each animal they have observed. This is an important part of scientific enquiry. We cannot base conclusions on one observation.

▲ 3 The students should choose their favourite fruit and do their own research into how the seeds of the fruit are dispersed. They can collect pictures and make their own drawings and then make these into a small poster in their Workbook on page 18.

Consolidate and review

- Ask students to make up a story or a poem about the journey of a seed from one place to another, with the help of an animal. Let some students read their stories out to the class afterwards.

Differentiation

■ All of the students should be able to name at least three animals that help to disperse seeds and describe how this happens in a simple way.

● Most of the students should be able to identify and observe (or conduct some basic research about) one or two animals in their local environment that help to disperse seeds. They should be able to make accurate notes about what the animals eat, and how and where the seeds are dispersed.

▲ Some of the students should be able to conduct their own research into how the seeds of a particular fruit are dispersed. They should be able to provide detailed and accurate information about seed dispersal.

1.9 What do seeds need?

Student's Book pages 18–19

Biology learning objective

- Investigate how seeds need water and warmth for germination, but not light.

Resources

- Workbook pages 19 and 20
- PCM B4: What do seeds need?
- Slideshow B7: Bean seedlings

Classroom equipment

- 16 bean seeds for each group, e.g. mung beans (soaked in water for 24 hours)
- transparent containers (jam jars)
- blotting paper or cotton wool
- water
- coloured pens or pencils
- rulers
- thermometers and measuring jugs or cylinders (optional)

Scientific enquiry skills

- *Plan investigative work:* Make predictions of what will happen based on scientific knowledge and understanding, and suggest and communicate how to test these; use knowledge and understanding to plan how to carry out a fair test; collect sufficient data to test an idea; identify factors that need to be taken into account in different contexts.
- *Obtain and present evidence:* Make relevant observations; measure volume, temperature, time, length and force.

Key words

- **germinate**
- **dormant**

Warn students not to eat any of the seeds, and to wash their hands after handling them.

Scientific background

Germination is the start of growth in the seed. Three factors are required for successful germination:

- water – this allows the seed to swell up and the embryo to start growing
- oxygen – so that energy can be released for germination
- warmth – germination improves as temperature rises (up to a maximum).

The seed absorbs water and a chain of chemical changes starts, which leads to the development of the young plant. Chemical energy stored in the form of starch is converted to sugar, which serves as food for the young plant during the germination process. Soon, the young plant is nourished and enlarged, and the seed coat bursts open. The growing plant emerges. The tip of the root emerges first, growing downwards, and helps to anchor the seed in place.

The seeds of some plants can lie *dormant* in the soil for many years until the conditions are right for germination to occur.

Students will be investigating what seeds need to germinate, focussing on water and warmth at this stage. Mung beans germinate quickly so they would be a good choice for these investigations.

Introduction

- Show the students Slideshow B7: Bean seedlings. Then ask: *What do these bean seeds need to start growing?* Write the sensible ideas on the board and explain that the students are going to conduct investigations to find out what the seeds need.
- Introduce the words 'germinate' and 'germination' and explain that they mean 'begin to grow'.

Teaching and learning activities

- Explain that the students are going to investigate what seeds need to germinate. Read the steps in the Student's Book with the class. Then discuss the way that the investigation needs to be set up, to help the students understand what is required in a fair test. Ask questions like: *Why do we need four jars? Why do we label the jars? Why do we need to put two jars in a sunny spot? Why do we put water in one jar and not in the other?*

- The students work in groups and set up their investigations. Walk around and make sure that they are doing this correctly. If you notice something incorrect, help the students to work out for themselves what they have done wrong. Remind them to carefully re-read the notes in their Student's Book.

Graded activities

As this is a practical investigation, it is recommended that the students work in mixed ability groups for this activity. Differentiation should be via the level of support the students receive as they work on the activity, as well as by outcome (please see guidance in the 'Differentiation' box right).

■1 Students discuss in their groups what they think will happen in each of the jars in the investigation they have set up. They should use the scientific knowledge that they already have of plants and not just make random guesses. Walk around and ask questions to develop their thinking, without giving them the correct answers. Each group should record their predictions about what will happen in each jar in the table on page 19 of their Workbooks. These can be shared with the rest of the class and later compared with the results of the investigation.

●2 After the groups have completed their investigations, the students should work individually to record what they observed in their Workbook on page 19. They also write answers to the questions in the Workbook, which will help them to clarify what they have observed and to draw conclusions.

After they have completed their observations and recordings, give the students copies of PCM B4 and let them answer the questions on this sheet. These questions will help them to think about the variables that they have tested, and whether or not they have conducted a fair test.

▲3 The students should make further investigations with the seeds that have germinated. They take measurements of the height (length) of the seedlings on six different days over the course of a few weeks. They record their measurements and comments about the growth of the seedlings in their Workbooks on page 20. Some more able students can also take further measurements regarding the temperature on each day and the amount of water given to each seedling.

Consolidate and review

- Ask each student to write a few sentences about what they investigated, what they predicted and what they found out.

Differentiation

■ All of the students should be able to set up a simple investigation, following step-by-step instructions and with some assistance. They should be able to predict what they think will happen and record their observations in a reasonably accurate way.

● Most of the students should be able take some basic measurements during the investigation, draw conclusions from their investigation and assess whether or not they have conducted a fair test.

▲ Some of the students should be able to continue their observations of the germinating seedlings and draw conclusions about the way in which germination conditions can affect the growth of seedlings. They should also be able to explain why it is important to conduct fair tests and understand ways to set about doing this.

1.10 Do seeds need light?

Student's Book pages 20–21

Biology learning objective

- Investigate how seeds need water and warmth for germination, but not light.

Resources

- Workbook page 21

Classroom equipment

- bean seeds for each group, e.g. mung beans (soaked in water for 24 hours)
- transparent containers (jam jars)
- blotting paper or cotton wool
- water
- rulers, thermometers and measuring jugs or cylinders

Scientific enquiry skills

- *Plan investigative work:* Make predictions of what will happen based on scientific knowledge and understanding, and suggest and communicate how to test these; use knowledge and understanding to plan how to carry out a fair test; collect sufficient data to test an idea; identify factors that need to be taken into account in different contexts.
- *Obtain and present evidence:* Make relevant observations; measure volume, temperature, time, length and force.

Key word

- **light**

Warn students not to eat any of the seeds, and to wash their hands after handling them.

Scientific background

Most seeds begin to germinate under the ground, where there is no *light*. Seeds do not need light to germinate, but seedlings will die if they do not get access to light quickly. Seeds need to be kept damp in order to germinate and they will die quickly if they have no access to water. Warmth is also a factor and seeds in any experiment should be kept at an even, warm temperature.

In this unit students will discuss and then set up their own fair tests to find out if seeds need light to germinate. They will also take accurate measurements in order to collect sufficient evidence to test an idea.

Introduction

- Start by asking students to look at the picture of a seed starting to grow in their Student's Book. Discuss where the seed starts to grow and prompt the students to think about the growing conditions under the soil. Ask: *Where does the seed germinate? Is it light or dark? Is it warm or cold? Is there water?*

Teaching and learning activities

- Explain to the students that they are going to set up their own investigations to find out if seeds need light to germinate. They will need to think about how they can set up a fair test. Discuss this as a class first, and remind students about variable factors. Explain that we need to have a control set of seeds so that we can compare the results.
- Students get into groups and discuss what they will need to do to conduct the investigation. They should discuss how many jars and seeds they will need, in what ways the germinating conditions should vary and where they will place in the seeds in the classroom, and then predict what will happen to the seeds.
- The students set up the investigation. Walk around and make sure that each group has set up their investigation correctly. Ask questions to develop the students' thinking.

Graded activities

As this is a practical investigation, it is recommended that the students work in mixed ability groups for this activity. Differentiation should be via the level of support the students receive as they work on the activity, as well as by outcome (please see guidance in the 'Differentiation' box right).

1 The students should discuss the investigation in their groups and suggest what they will do to make it a fair test. They should also predict what will happen to the seeds. They should write about their investigation and predictions on page 21 of their Workbook.

2 The students should record their observations over a period of seven days. This time they will need to draw up their own table for this. They should look at other tables they have completed in their Workbook for ideas about how to do this. Encourage them to record measurements of height, volume of water, hours of sunlight and temperature, in order to collect as much data as possible.

3 The students should write a paragraph in which they compare what things seeds and plants need to grow. They should apply what they know about seeds and plants. For example, seeds have a food supply when they start to grow. Plants make their own food to grow and, to do this, they need light. Students should be able to refer back to the process of photosynthesis and make connections.

Consolidate and review

- Students can work in pairs. Each student draws up five questions to ask his or her partner about what they have learned. They should each write down the questions and their expected answers. Walk around, checking that the questions are appropriate and the answers correct.

Differentiation

■ All of the students should be able to set up a simple investigation with some assistance and have some awareness of what makes a fair test. They should be able to predict what may happen and record their observations in a reasonably accurate way. All of the students should be able to state briefly what seeds need to grow.

● Most of the students should be able to set up an investigation with little assistance, decide how to record their observations, draw conclusions from their investigation and assess whether or not they have conducted a fair test. They should also be able to write a short paragraph in which they state what seeds need to germinate and what plants need to grow.

▲ Some of the students should be able to work quite independently to set up a fair investigation and record their observations. They should be able to take measurements and draw conclusions. They should also be able to write a paragraph comparing the growing needs of seeds and plants, and referring to their own scientific knowledge about plants.

1.11 Growing seeds in variable conditions

Student's Book pages 22–23

Biology learning objective

- Investigate how seeds need water and warmth for germination, but not light.

Resources

- Workbook page 22

Classroom equipment

- bean seeds or other fast-growing seeds for each group
- transparent containers (jam jars)
- blotting paper or cotton wool
- water
- rulers, thermometers and measuring jugs or cylinders

Scientific enquiry skills

- *Plan investigative work:* Make predictions of what will happen based on scientific knowledge and understanding, and suggest and communicate how to test these; use knowledge and understanding to plan how to carry out a fair test; collect sufficient data to test an idea; identify factors that need to be taken into account in different contexts.
- *Obtain and present evidence:* Make relevant observations; measure volume, temperature, time, length and force.
- *Consider evidence and approach:* Interpret data and think about whether it is sufficient to draw conclusions.

Warn students not to eat any of the seeds, and to wash their hands after handling them.

Scientific background

Seeds require water, oxygen and warmth to germinate successfully. Small plants also require light so that they can start making their own food in order to grow.

In this unit, students will measure how seeds grow in variable conditions. To do this they will control and measure the temperature and the volume of water given to each seed, and then measure its growth over a period of two weeks.

Introduction

- Discuss the investigations that have already been done and let the students summarise what seeds need in order to germinate and grow.

Teaching and learning activities

- Explain to students that they are going to investigate how well seeds grow in variable (different) conditions. Ask them which growing conditions they could vary. Talk about plants in the garden and the ways that they react in different growing conditions. Ask: *Do plants grow faster when it is hot or cold? Do they grow faster when it rains or when it doesn't rain?*

Graded activities

As this is a practical investigation, it is recommended that the students work in mixed ability groups for this activity. Differentiation should be via the level of support the students receive as they work on the activity, as well as by outcome (please see guidance in the 'Differentiation' box right).

1 Students get into groups and discuss what they will need to do to conduct the investigation. They should discuss how many jars and seeds they will need to conduct a fair test and then discuss in what ways the germinating and growing conditions should vary. They also need to discuss how they will take measurements and how often they will do this. Over a given time, they need to measure the volume of water given to each seed, the height of each plant and the temperature.

The students should set up their investigation. As they do so, walk around and make sure that each group has set up their investigation correctly. Ask questions to develop their thinking. Make sure that all the students in the group are involved in setting up the investigation.

2 All the students should record the growth of their seeds and record the measurements they make in their Workbook on page 22. They will have to share out the work, but should all copy the measurements accurately into their own Workbooks. Stress that accuracy is very important when collecting scientific data.

3 In their groups, the students should draw conclusions from their investigations and then take part in a class feedback session. Each group should present their conclusions, explaining what data they used to reach these conclusions. If, for example, they concluded that seeds need a certain amount of water to grow well, they will need to illustrate this with the measurements they took about the heights of the seedlings that did germinate and grow.

Consolidate and review

- Students can work alone or in pairs and research what happens to seeds that lie dormant, for example during a drought. They can describe what conditions the seeds need to start germinating.

Differentiation

■ All of the students should be able to take part in an investigation, take basic measurements and record their measurements.

● Most of the students should be able to set up an investigation with little assistance, decide how to record their observations, draw conclusions from their investigation and assess whether or not they have conducted a fair test.

▲ Some of the students will be able to work quite independently to set up a fair investigation and record their observations. They should be able to take measurements and draw conclusions.

1.12 The life cycle of a plant

Student's Book pages 24–25

Biology learning objective

- Recognise that flowering plants have a life cycle including pollination, fertilisation, seed production, seed dispersal and germination.

Resources

- Workbook pages 23 and 24
- PCM B5: The life cycle of a tomato
- Slideshow B3: Flowering plant life cycle

Classroom equipment

- posters illustrating the life cycles of plants (other than a bean or sunflower)
- additional pictures of young plants
- coloured pens or pencils
- large sheets of paper to make posters

Scientific enquiry skills

- *Plan investigative work:* Identify factors that need to be taken into account in different contexts.
- *Obtain and present evidence:* Make relevant observations.

Key word

- **life cycle**

Scientific background

Flowering plants grow from seeds, which are produced by parent plants after pollination and fertilisation have taken place. The seeds develop inside fruits, which are formed from the flowers of the plant. The flowers grow on a plant that has, itself, grown from a seed. The seeds are dispersed, and they germinate when conditions are favourable. A seed absorbs large quantities of water; the plant embryo swells until the seed coat bursts and the tip of the root of the new plant emerges. This root anchors the seed in place and enables the embryo to absorb nutrients and water from the soil. The plant begins to grow and eventually matures, before entering a new cycle of growth. This *life cycle* is continuous, with plants producing seeds that will produce new plants.

Introduction

- Show Slideshow B3 again about the flowering plant life cycle. Let the students describe the different stages in the life of a plant that they have learned about: pollination, fertilisation, seed production, germination, and seed dispersal.
- Then talk about cycle diagrams. Show the students different examples and look at the examples in the Student's Book. Discuss why we use cycle diagrams and what they illustrate about life.

Teaching and learning activities

- Let students compare the linear diagram and the cycle diagrams in their Student's Book, and discuss the differences. They should compare the information in the two types of diagrams and identify the stages of plant life shown in each. The only difference is that the linear diagram does not show that plant life happens in a continuous cycle.
- Students study the life cycles of a bean and a sunflower, as shown in the diagrams in their Student's Book. Discuss each diagram and let the students describe each stage in as much detail as possible. Ask questions like: *Which picture shows germination? Which picture shows pollination? Where do the seeds develop? How do you think the seeds are dispersed? What do the seeds look like? What do bean flowers look like? What do sunflowers look like?*
- Let students work in pairs and describe the life cycles to each other.
- If you have made or collected any other plant cycle diagrams, show them to the class and discuss what we can learn about each plant from the diagram.
- Then compare the life cycles of the two plants. Ask questions like: *Are the seeds of the plants the same or different? Do seed dispersal and pollination occur in the same way for both plants? Are the flowers of the plants similar or different?*

Graded activities

1 The students should label the stages in the life of a plant in their Workbook on page 23.

2 Students draw a life cycle of a sunflower in their Workbook on page 24. They should use the information given in their Student's Book and transfer this to their own diagrams. They will need to draw pictures and write a short description of each stage.

3 The students should choose a plant that they are familiar with and make a poster to illustrate its life cycle. If they need help choosing a plant, suggest that they draw the life cycle of a fruit tree. They should include a sentence describing how the seeds are dispersed (by wind, water, animal, etc.).

Consolidate and review

- Make copies of PCM B5: The life cycle of a tomato. Cut the pictures up and place them in small envelopes or bags. Let students work in pairs. They should put the pictures in order and then use the information in the pictures to draw a life cycle diagram. They should label each stage of the diagram.

Differentiation

■ All of the students should be able to identify and name the main stages in the life of a plant and recognise that the life of a plant is a cycle.

● Most of the students should be able to create a life cycle diagram using information from another source (transfer the information) with some assistance. Most students should also be able to read and interpret a variety of simple life cycle diagrams.

▲ Some of the students should be able to carry out further research about the life cycle of a plant of their choice. They will create informative and accurate posters and will be able to say how the seeds of their plant are dispersed.

Any students who have read *Big Cat Fly* facts may be aware that flies are among the insects that pollinate plants, which is a vital part of the plant's life cycle.

1.13 Stages in the life cycle

Student's Book pages 26–27

Biology learning objective

- Recognise that flowering plants have a life cycle including pollination, fertilisation, seed production, seed dispersal and germination.

Resources

- Workbook pages 25 and 26
- DVD Activity B2: Plant life cycle

Classroom equipment

- pictures and posters of plant life cycles
- collection of different types of seeds
- plants grown in class by the students
- fresh flowers
- rulers
- large sheet of paper for class bar chart
- colouring pens or pencils

Scientific enquiry skills

- *Obtain and present evidence:* Make relevant observations; present results in bar charts and line graphs.

Key words

- **life cycle**
- **pollinated**
- **fertilised**
- **ovum (*plural* ova)**
- **seeds**
- **germinate**

> ⚠ Check that none of the seeds are poisonous. Warn students not to eat any of the flowers, plants, fruit or seeds and to wash their hands after handling them. Check that none of the seeds are poisonous. Be aware that some students may suffer from hay fever, and take necessary precautions.

Scientific background

This unit revises and helps students to consolidate what they have learned about plants and their *life cycles*: *germination*, growth, reproduction (*pollination*, *fertilisation*, *seed* formation), seed dispersal.

Each stage in the life cycle of a plant is very important. If the cycle is broken, the plant will not reproduce. Eventually, the plant will die. If this happens to all the plants in a species, the species will become extinct.

Introduction

- Refer to a life cycle chart and ask questions like: *What would happen if there were no animals to disperse seeds? What would happen if the weather changed and it was always dark and cold? What would happen if the average annual temperature rose?* Students should refer to the cycle chart as they attempt to answer the questions.

Teaching and learning activities

- Start by looking at some of the real seeds you have collected and the plants that the students have grown in class. You could also bring some fresh flowers to the class. Students can relate these to the stages on a life cycle diagram.
- Read the information table in the Student's Book with the students and discuss it. The table provides a short summary of the key stages in the life cycle of a flowering plant.
- Ask students if they can add anything to the table. Ask: *Who or what helps to disperse seeds?*
- Then let students work in pairs to ask each other questions about the table, to consolidate what they have learned. Observe the students as they work, and ask questions yourself to check their understanding. Guide them to correct answers but don't give them the answers directly.
- Show any of the slides or videos that you have shown while teaching this topic. Revise any section of the work that students do not seem to have understood well.
- Allow students to complete DVD Activity B2 to remind them of the stages in a plant life cycle.

Graded activities

■1 In groups, the students should discuss what they think would happen if no seeds from one particular type of plant germinated. Prompt them with questions such as: *Could this happen? Why?*

●2 The students should draw a life cycle diagram using the key words given in their Workbook on page 25. After they have completed the diagram, they should write a short explanation of each stage. They can refer to previous life cycle diagrams for information when they do this. They can also refer to the glossary at the back of their Student's Book or to the relevant pages in their Student's Book. Help students to use the resources that they have in their books to compile and check their own answers, if they do not remember everything. Being able to find the correct information is as important as being able to remember it.

▲3 The students should, in the same groups as they were in for Unit 1.11, take the final measurement of the tallest plant from their investigation. On a large class poster, the students should draw a bar chart to show their result. The students can then see which group has grown the tallest plant. They should ask questions such as: *What were the conditions that produced the tallest/shortest plant? Are all the plants similar heights? Why?*

Consolidate and review

- Students can fill in the gaps on page 26 of the Workbook to revise what they have learned. Give them the answers afterwards and let them check their own answers.

Differentiation

■ All of the students should be able to suggest what might happen if all of the seeds from one particular type of plant do not germinate. Accept all reasonable answers.

● Most of the students should be familiar enough with life cycle diagrams to be able to create a life cycle diagram using information from another source (transfer the information). Most students should also be able to read and interpret a variety of simple life cycle diagrams.

▲ Some of the students should be able to measure their plants and create an accurate bar chart of the results. They should conclude that the seeds with the best growing conditions grew the best.

Consolidation

Student's Book page 28

Biology learning objectives

- Know that plants need energy from light for growth.
- Know that plants reproduce.
- Observe how seeds can be dispersed in a variety of ways.
- Investigate how seeds need water and warmth for germination, but not light.
- Know that insects pollinate some flowers.
- Observe that plants produce flowers which have male and female organs; seeds are formed when pollen from the male organ fertilises the ovum (female).
- Recognise that flowering plants have a life cycle including pollination, fertilisation, seed production, seed dispersal and germination.

Resources

- Assessment Sheets B1, B2, B3 and B4

Looking back

Use the summary points in the Student's Book to review the key things that the students have learned in this topic. Ask suitable questions to stimulate discussion.

Ask students to write down five questions about things they have learned in this topic. Ask students to exchange questions and answer each other's. They should check the answers themselves.

How well do you remember?

You may use the revision and consolidation activities on Student's Book page 28 either as a test, or as a paired class activity. If you are using the activities as a test, have the students work on their own to complete the tasks in writing. Collect and mark the work. If you are using them as a class activity, you may prefer to let the students do the tasks orally, in pairs. Circulate as they discuss and answer the questions; observe the students carefully to see who is confident and who is unsure of the concepts.

Some suggested answers

1 Answers will vary. Students should draw a picture similar to the one on page 9 of their Student's Book.

2 Students should make the following points:
  - Plants need sunlight, air and water to make their food.
  - The roots of the plants take up water from the soil. The water moves up through the stem to the leaves.

3 The bee picks up pollen from the anthers of one flower and the pollen brushes off and falls on the stigma of another flower as the bee moves from flower to flower. Pollination is an essential step in the plant's life cycle. If no pollen reaches the stigma, the flower will not reproduce and no new plants will grow.

4 Students' own drawings.

Assessment

A more formal assessment of the students' understanding of the topic can be undertaken using Assessment Sheets B1, B2, B3 and B4. These can be completed in class or as a homework task.

Students following Cambridge International Examinations Primary Science Curriculum Framework will write progression tests set and supplied by Cambridge at this level, and feedback will be given regarding their achievement levels.

Assessment Sheet answers

Sheet B1

1 false / true / true [3]

2 anthers / insects / stigma / fertilisation [4]

3 The plant life cycle will not continue if a plant does not reproduce. If all the plants in one area fail to reproduce, that type of plant will die out. [1]

4 false / true [1]

5 dispersal / growth [2]

6 Students' own answers, e.g. a bee. It moves around to many plants because it is attracted by the flowers. The pollen sticks to its body and falls off on other flowers. [2]

7 seed / pollination [2]

Sheet B2

1 The ovary. [1]

2 true / false [1]

3 the anthers [2]

4 ovary / stamen or anther [2]

5 Plants need light [1] and water [1] to grow well. Seeds need warmth [1] and water [1] to grow well.

6 water / animals / wind / explosion [4]

7 light [1]

Sheet B3

1 Female parts: carpel, ovary, ovum, stigma, style.
Male parts: stamen, filament, anther. [8]

2 wind / exploding / animals / water [4]

3 Students' own diagrams. [3]

Sheet B4

1 false / true / true / false / true / true [6]

2 Pollen is made, by the anthers. The colourful petals of a flower, attract insects. Bees are good, pollinators. The fertilised part of the flower, develops into a fruit. [4]

3 pollen / hooks / bodies / stigma / stigma [5]

Student's Book answers

Pages 2–3

1 Answers might include: leaf, stem, flower, fruit, roots.

2 Plants needs leaves to make food; they need roots to anchor them in the ground and take up water from the soil; they need stems to support the plant; they need flowers to grow into fruits and seeds.

3 water, light, soil

Pages 4–5

1 Water them regularly, let them have plenty of sunlight, do not let them get too cold.

2 Plant A has green healthy leaves. Plant B has yellow/brown wilted leaves. If the leaves do not get enough light they cannot make food for the plant.

3 Students discuss how a plant makes its own food in pairs.

Pages 6–7

1 The plantlets have leaves, stems and roots. The runners grow along the surface of ground.

2 The fruits (or the flowers).

Pages 8–9

1 Male parts: stamen (anther, filament).
Female parts: carpel (ovary, ovum, stigma, style).

2 Students' own answers.

Pages 10–11

1 Students should be able to identify these stages: anthers make pollen; pollen lands on the stigma; tubes grow down the style; pollen goes into the ovary; pollen joins with the ovum; seeds grow inside the ovary.

2 Seeds grow inside the ovary. They have a hard coating that protects them.

Pages 12–13

1 The flowers are brightly coloured and some smell nice to insects. The insects pollinate the flowers.

2 Students' own answers.

Pages 14–15

1 Students' own answers.

2 Seeds are dispersed by water, by wind, by animals, by the seed pods exploding.

Pages 16–17

1 Students' own answers.

2 The burrs are being carried away on the horse's mane to a different place.

3 The seeds may grow into new plants.

Pages 18–19

1 Students' own answers.

2 Students' own answers, e.g. to produce more than one set of results, in case some of the seeds do not germinate.

Pages 20–21

1 Under the ground / dark / warm / yes
it grows a stem / inside the seed case / light.

Pages 22–23

1 The farmer can control the conditions inside the greenhouse and so help the plants to grow well in unsuitable areas. The farmer can control water and temperature (and light)

2 When it is warm.

Pages 24–25

1 diagram A / the life cycle starting again / diagram A

2 Students' own answers.

Pages 26–7

1 Students' own answers.

2 Students' own answers, e.g. photosynthesis takes place, how seeds are dispersed.

2.1 Solids, liquids and gases

Student's Book pages 30–31

Chemistry learning objective

- (Revise Stage 4 objective) Know that matter can be solid, liquid or gas.

Resources

- Workbook pages 27, 28 and 29
- Video C1: Solids, liquids and gases
- DVD Activity C1: Solids and liquids

Classroom equipment

- selection of materials, especially water, in different states for students to observe and test
- a fridge and containers for freezing water, if possible
- a kettle, or a burner and small pan

Scientific enquiry skills

- *Plan investigative work:* Make predictions of what will happen based on scientific knowledge and understanding, and suggest and communicate how to test these.
- *Obtain and present evidence:* Make relevant observations.

Key words

- **solid**
- **liquid**
- **gas**
- **states of matter**
- **properties**
- **temperature**

⚠ The students should be kept back when water is being boiled and should not handle containers of hot water. Warn the students that spillages will make the floor slippery. Make sure all spillages are cleaned up promptly.

Scientific background

Solid, *liquid* and *gas* are the three main *states of matter*. These are the normal forms of matter on Earth: nearly all materials can take these three forms. The three states are characterised by different *properties*.

Solids are characterised by their fixed shape, and their inability to flow or to be compressed.

Liquids take the shape of any container, can flow, have a surface and cannot be compressed easily. The measure of the amount of a liquid is called its volume and the volume stays the same, even though the shape of the liquid can change in different containers.

Gases also take the shape of any container and can flow. They can be compressed easily. Unlike liquids, they do not have a surface, and they expand to fill their container.

In solids, the particles are held together in a regular pattern by relatively strong forces, but vibrate about a fixed point. In liquids, the forces between the particles are weaker and the particles can move from place to place. The forces keeping gas particles together are extremely weak; the particles are much further apart and move about very quickly. This is why gases are compressible.

Introduction

- Show the students a block of ice, a bowl of water and steam coming from a kettle. Ask: *Which one of these is a solid? Which is a liquid? Which is a gas?* Write the words 'solid', 'liquid' and 'gas' on the board. Remind the students that matter exists in different states. Ask the students for some other examples of solids, liquids and gases.
- Play Video C1: Solids, liquids and gases and discuss what the students can see.
- Let the students complete DVD Activity C1 to remind them about the properties of solids and liquids.

Teaching and learning activities

- Let students look at the pictures in their Student's Book. Read and discuss the captions under the pictures. Remind the students that matter has certain properties like having a shape, taking the shape of its container and being able to be poured.
- The students then look at the pictures of water on pages 29 and 31 in their Student's Book. They should discuss the properties of water in each state shown in the pictures.

- Introduce the idea that water can change state. Ask: *Can water change from one from state to another?* Let the students give examples.
- If you have a fridge or freezer near the classroom, bring some ice to the class and let the class watch it melt. Put some ice in a warm spot and some ice in a cooler spot and ask students to predict what will happen. Then observe how fast the ice changes state in each spot. The students should be able to conclude that ice melts faster when the temperature is higher.
- Put some water in small plastic bags or small flat containers. Ask: *What can we do to change the water from a liquid state to a solid state? How long will this take?* Let the students experiment by freezing water in different types of containers.
- Then discuss how liquid or solid water can be changed into a gas. If you have a kettle, measure an amount of water and put the water in the kettle. Let it boil. Or measure some water, put it in a small pan and boil the water over a burner for 10 minutes. Let the water cool and then measure the amount of water again. The volume should be lower because some of the water has changed into water vapour (a gas). Ask: *What caused the water to change from a liquid to a gas?*

Graded activities

It is recommended that the students work in mixed ability groups for these revision activities. Differentiation should be via the level of support the students receive as they work on the activity, as well as by outcome (please see guidance in the 'Differentiation' box right).

1 The students should work alone to complete the sentences on page 27 of their Workbooks.

2 The students should revise the three states of matter by observing and testing some materials. Provide a selection of materials that the students can squeeze, pour, spray and touch. Keep the focus on water in different states but include a few other materials as well. Students can record their observations in their Workbooks on page 28.

3 In groups, the students should develop their understanding of what they observed during the testing of the materials by discussing the questions in the Student's Book.

Consolidate and review

Sum up by concluding that:

- solids have a fixed shape, cannot flow and cannot be squashed
- liquids take the shape of any container, can flow and cannot be squashed
- gases take the shape of any container, can flow and can be squashed
- water exists in three shapes: as a solid (ice), a liquid (water that we can pour) and a gas (water vapour).

Students can complete Workbook page 29 to consolidate and review what they know about the states of matter and to identify the three states of water.

Differentiation

■ All of the students should be able to complete the revision exercise in their Workbook with little or no help.

● Most of the students should be able to test the different materials by touching, pouring and observing in order to decide a material's state. They should be able to record their results with little or no help.

▲ Some of the students should be able to discuss what happens when matter changes state.

2.2 Liquid to gas – evaporation

Student's Book pages 32–33

Chemistry learning objective

- Know that evaporation occurs when a liquid turns into a gas.

Resources

- Workbook pages 30–31 and 32–33
- DVD Activity C2: Heat and evaporation

Classroom equipment

- spray perfume and shallow dish
- for each group: 3 T-shirts (or identical pieces of fabric), water, clothesline and pegs, weighing scales
- for each group: three open containers of different diameters, measuring cylinder or jug, water

Scientific enquiry skills

- *Obtain and present evidence:* Present results in bar charts and line graphs.
- *Plan investigative work:* Make predictions of what will happen based on scientific knowledge and understanding, and suggest and communicate how to test these; use knowledge and understanding to plan how to carry out a fair test; identify factors that need to be taken into account in different contexts.
- *Consider evidence and approach:* Decide whether results support predictions; interpret data and think about whether it is sufficient to draw conclusions.

Key words

- **evaporate**
- **surface area**
- **rate**

Warn the students that spillages will make the floor slippery. Make sure all spillages are cleaned up promptly.

Scientific background

Evaporation is a change of state in which a liquid changes to a gas. This can happen when liquids are cold or hot. Evaporation is greater and faster when the temperature is higher, because the liquid particles have more energy and are moving faster. The likelihood that a particle will escape, or evaporate, from the liquid is greater. Evaporation only takes place from the surface of the liquid, where the particles are able to escape into the air and become a gas. The bigger the *surface area*, the greater the evaporation. Movement of air over the surface of a liquid will also increase the *rate* of evaporation. A high rate of evaporation means that wet things dry quickly. A low rate means that wet things dry slowly. If there is a large surface area, with lots of air movement, and the temperature is high, then evaporation will happen quickly. If the temperature is low, the air is still and the surface area is small, then the rate of evaporation will be slow.

Introduction

- Spray some perfume into a shallow dish. It will quickly appear to disappear. Ask: *Where did it go? What state of matter has the perfume changed to? Why did the perfume change from a liquid to a gas?*
- Explain that heat is needed for a liquid to become a gas, but sometimes the heat in the environment is enough. Introduce the word 'evaporation' as the term used when a liquid changes to a gas. Students should be familiar with the term from earlier work.
- Let the students complete DVD Activity C2 to remind them what they learned in Stage 4.

Teaching and learning activities

- Ask the students to look at the picture of the puddle on page 32 of the Student's Book. Answer the questions as a class.
- Make sure that the students understand that evaporation from a water surface can happen even at everyday temperatures. Ask: *So what is the effect of heat?* (Liquids evaporate faster at higher temperatures.)
- Ask the students to look at the other pictures on pages 32 and 33 of the Student's Book. Discuss the ways in which temperature, air flow and surface area can affect evaporation.
- Ask the students to think about the following scenarios (select different students to respond): *How much water would remain if you left a glass of water out at night? A glass of water on a hot stove?*

A glass of water in the sun? What could you do to make this a fair test?

Graded activities

1 The students should do a practical activity outside. They should use the same amount of water to create different puddles in the same sunny area. One puddle should be shallow, another deep; one should have a larger surface than the other. Ask: *Which puddle will dry up first? Why?* Leave the puddles for a few hours, or time as appropriate, and then let the students observe what has happened. Was their prediction correct?

2 The students should work in groups to set up their own investigations to find what other factors affect the rate at which wet clothes dry. Ask the groups to plan an investigation to test their ideas. Take feedback from the groups and review proposals. Establish whether their ideas will provide a fair test. Ask the students to make some predictions about which clothes will dry the quickest. Students can record the results of their investigations in their Workbook on pages 30–31. They should read the pages carefully and answer all the questions there.

3 The students should work in groups to set up their own investigations to find out if the shape of a container (and, therefore, the surface area of the water in it) can affect how quickly water evaporates. Ask the groups to plan an investigation to test their ideas. Take feedback from the groups and review proposals. Establish whether their ideas will provide a fair test. Ask the students to make some predictions about which container the water will evaporate from the fastest. Students can record the results of their investigations in their Workbook on pages 32–33. They should read the pages carefully and answer all the questions there.

Consolidate and review

Ask the groups to report back on the investigations that they carried out. They should be able to explain what they investigated, what they did to make it a fair test, what they predicted, and what the results showed.

Differentiation

■ All of the students should be able to take part in the puddle investigation. They should be able to predict that the shallowest puddle with the greatest surface area will evaporate fastest. They should also be able to describe evaporation as a change of state which can occur at any temperature.

● Most of the students will be able to set up their own investigation into factors that will cause clothes to dry more quickly. They should be able to plan the investigation, set it up, predict the results and record their measurements in tables and bar charts with little assistance.

▲ Some of the students should be able to set up and participate in a group investigation to test how surface area affects evaporation. They should be able to predict that the container with the largest surface area will cause the fastest evaporation. They should be able to record their results and draw a simple bar chart of the results. More able students should be able to describe the process of evaporation, say when it is faster or slower, and be able to explain why the greatest evaporation occurs when it is hot.

2.3 Gas to liquid – condensation

Student's Book pages 34–35

Chemistry learning objective

- Know that condensation occurs when a gas turns into a liquid and that it is the reverse of evaporation.

Resources

- Workbook pages 34 and 35

Classroom equipment

- kettle, or saucepan and heat source
- water
- mirror or tile
- oven gloves
- two plastic cups, one slightly bigger and taller than the other, for each group of students
- hand lenses
- ice cubes
- large sheets of paper to make posters
- coloured pens or pencils

Scientific enquiry skills

- *Ideas and evidence:* Use observation and measurement to test predictions and make links.
- *Plan investigative work:* Make predictions of what will happen based on scientific knowledge and understanding, and suggest and communicate how to test these; use knowledge and understanding to plan how to carry out a fair test; collect sufficient evidence to test an idea; identify factors that need to be taken into account in different contexts.
- *Consider evidence and approach:* Decide whether results support predictions.

Key words

- **condense**
- **reverse**

⚠ The students should be kept back when water is being boiled and should not handle containers of hot water. Warn the students that spillages will make the floor slippery. Make sure all spillages are cleaned up promptly.

Scientific background

Gas is one of the three states of matter. Liquids can turn to gases and gases can change back to liquids. These changes are the *reverse* of each other. Gases can change back to liquids when cooled. When the heat energy decreases, the gas particles lose energy, slow down and come closer together to make droplets of liquid. This is called *condensing*. It can happen at any temperature between the melting point and the boiling point of the material, but is faster at lower temperatures.

Introduction

- Ask the students: *Have you noticed what happens to mirrors in steamy rooms or saucepan lids when cooking?*
- Then do a demonstration. Boil some water and let the steam (water vapour) hit the cold surface of a tile or mirror. Hold the mirror or tile with oven gloves. Warn the students about the dangers of steam – it is extremely hot and can cause blistering or burns. Discuss the beads of water that appear on the cold surface. Ask: *What is happening?* Explain that this is condensation. The water vapour (gas) is hitting the cold mirror or tile, cooling and changing state back to liquid water.

Teaching and learning activities

- Ask the students to look at the experiment pictured on page 35 of the Student's Book. Encourage students to ask questions about the investigation. Ask: *What do you think will happen?*
- Ask students, in groups, to set up the investigation on page 35 of the Student's Book. They should carefully fill the bottom glass about two-thirds full of warm water. Then they should invert the taller glass and place it on top of the glass with the water. Ask: *What do you think will happen? Where? Why?* Lead them to understand that some of the warm water in the bottom glass will evaporate and water vapour will be formed. This water vapour will rise and reach the cooler air in the top glass. The cooler air will

cause the water vapour to condense and form little droplets of water.

- Tell the students to wait for a few minutes and then give them a hand lens. Tell them to look carefully at the top glass. Ask: *What do you observe? Where are the droplets of water? Where did the water droplet come from?* Encourage students to use the words 'condense' and 'condensation'.
- Then ask: *What conditions will affect this investigation? How can we increase the condensation?* Allocate some students to investigate the effect of putting ice on top of the top glass to see if this affects the rate and amount of condensation. They should use two separate sets of glasses to test this, one with ice and one without ice. Other students can investigate the effect of the temperature of the water on condensation. They could use two sets of glasses and put cold water in one bottom glass and warm/hot water in the other bottom glass. Establish what students could do to make each investigation a fair test.
- Students should leave their glasses for about 3-5 minutes and then make their comparisons. Ask: *Which glasses produced the most condensation? Why do you think this is?*

Graded activities

■ 1 The students should work in pairs. Each student draws a diagram and uses the diagram to explain to their partner why condensation is the reverse of evaporation.

● 2 The students should work in groups and set up an investigation to find out what happens to ice when it is warmed. Ask the groups to plan an investigation to test their ideas. Establish whether their ideas will provide a fair test. Ask the students to make some predictions. Students can record the results of their investigation in their Workbook on page 34. They should read the page carefully and answer all the questions there.

▲ 3 The students should discuss the picture in the Workbook on page 35. Discuss the questions with the students and help them to understand that evaporation and condensation are being used to collect water in a dry place. The students can then make their own poster to show how it is possible to collect water in a very dry place.

Consolidate and review

- Use a diagram to revise and consolidate the idea that condensation is the reverse of evaporation.

Differentiation

■ All of the students should be able to give everyday examples of condensation and recognise how and when it occurs.

● Most of the students should be able to describe condensation as a reversible change, when a gas is cooled to a liquid.

▲ Some of the students should be able to explain where the water comes from in different situations and describe how it has condensed. They can plan a fair test to see which factors affect condensation the most.

2.4 Water vapour in the air

Student's Book pages 36–37

Chemistry learning objective

- Know that air contains water vapour and when this meets a cold surface it may condense.

Resources

- Image C1: The water cycle

Classroom equipment

- photographs of water in different states
- access to the internet or reference books
- presentation software
- sheets of paper to make posters
- coloured pens or pencils

Scientific enquiry skills

- *Obtain and present evidence:* Make relevant observations.

Key words

- **water cycle**
- **reversible**

Scientific background

Water is a liquid at everyday temperatures: it can change state from a liquid to a gas (water vapour). If water is frozen, it changes its state to a solid (ice). These changes are *reversible*: the water vapour can be condensed to form water droplets, and the ice can be melted to form liquid water. There is always water vapour in the air.

These changes are evident in the *water cycle*. Water is essential to all life in Earth.

Please note that the students do not need to learn about the water cycle for the Cambridge Progression or Primary Checkpoint tests. This material is included here in addition to the objectives stated by the Cambridge Primary Science curriculum framework.

Introduction

- Use this unit to establish the importance of water to life on Earth and to introduce students to the water cycle. This will help them to understand the importance of evaporation and condensation and that one process is the reverse of the other process.

Teaching and learning activities

- Start by talking about the importance of water. Ask students for examples of ways in which we use water in different states. Make sure they understand that snow, ice, hail and rain are all forms of water. Ask: *What happens if we don't have enough water? Where has the water gone?*
- Remind the students of the terms 'evaporation' and 'condensation' and encourage them to give examples of each. The students should now have discovered that there is always water vapour in the air, which can condense as it cools. They should also know that water on the surface of the Earth (for example, in puddles) evaporates into the air to form water vapour. Reinforce the fact that evaporation and condensation are the reverse of each other.
- Use the pictures in the Student's Book and any other pictures that you may have. Let students describe the state of water in each picture and say what could cause the water to change its state.
- Then introduce the idea that water goes around and around in a cycle on Earth – it is never lost. Show the students Image C1: The water cycle. There may be more water in some places and less water in other places, but there is always water on Earth. Most of the water is in the sea.

Graded activities

1 Ask students to find out some interesting facts about water or about rain. Here are some ideas: Water is made up of hydrogen and oxygen and has the chemical formula H_2O; more than 70% of the surface of the Earth is covered in water; pure water has no taste or smell; rainforests are forests that get a lot of rain; Antarctica is the place on Earth where the least amount of rain falls.

The students should produce a slide presentation or storyboard of their facts to present to the rest of the class. Encourage them to use facts about water in its different states on Earth.

2 The students should make lists of places on Earth (or places in their country) from which water can evaporate. Encourage them to think about rivers, oceans, reservoirs, swimming pools, baths, puddles, glasses of water, etc.

3 The students should work in pairs and describe to their partners where rain comes from. Some students may have started to understand the processes of the water cycle. Let them use pictures or diagrams in their explanations. Walk around, assisting students as necessary.

Consolidate and review

- If time allows, the students could design posters on why we should be very careful with our water and why we should not waste or poison it.
- Students could also study the rainfall in their own country and identify the places that get the most and the least amounts of rain each year.

Differentiation

■ All of the students should be able to find out some interesting facts about water on Earth using the internet or reference books. They will be able to present their facts to the class in an interesting and informative way.

● Most of the students should be able to identify the places on Earth from which water evaporates.

▲ Some of the students should be able to explain how rain is formed and that this is a cyclical process. More able students may be able to start drawing a diagram to show this cycle.

2.5 The water cycle

Student's Book pages 38–39

Chemistry learning objectives

- Know that condensation occurs when a gas turns into a liquid and that it is the reverse of evaporation.
- Know that air contains water vapour and when this meets a cold surface it may condense.

Resources

- Workbook pages 36 and 37
- DVD Activity C3: The water cycle

Classroom equipment

- poster of the water cycle, if available
- sheets of paper and other materials to make a collage (optional)
- to make a model of the water cycle, each group will need: a large glass or plastic bottle with a lid, some small pebbles, soil and sand, a small green plant that will fit inside the jar, a small container for water
- coloured pens or pencils

Scientific enquiry skills

- *Ideas and evidence:* Use observation and measurement to test predictions and make links.
- *Obtain and present evidence:* Make relevant observations.

Key word

- **water cycle**

Warn the students that spillages will make the floor slippery. Make sure all spillages are cleaned up promptly.

Scientific background

Water falls as precipitation, either solid precipitation (snow or hail) or liquid (rain). This water reaches the rivers and, eventually, the sea. Virtually all the world's water at any one time is in the sea. The water at the surface of the sea evaporates to form water vapour. Some of this water vapour rises and comes into contact with cooler air higher up. Consequently, the water vapour cools and some of it returns to the liquid state, forming small water droplets. This is what we see as clouds (aircraft condensation trails are a similar phenomenon). Through further cooling and other effects, the water droplets in the clouds may become larger until they are heavy enough to fall as precipitation. The process then begins again, forming the *water cycle*.

Please note that the students do not need to learn about the water cycle for the Cambridge Progression or Primary Checkpoint tests. This material is included here in addition to the objectives stated by the Cambridge Primary Science curriculum framework.

Introduction

- Show the students the poster of the water cycle, if you have one, or refer to the diagram in the Student's Book on pages 38–39. Give them a few minutes to study the diagram and then ask them what they think it shows. Prompt them with questions like: *Why is there an arrow here? Where is the water going? Why is it changing state here?*

Teaching and learning activities

- Using the poster or the diagram in the Student's Book, discuss each stage of the water cycle with the class. Ask students to identify the state of water in each stage, and then to name the change of state when it moves to another stage. Here are some of the points that you should explain:
- Rain falls and collects in streams and rivers. This is water in a liquid state. Snow falls on high mountains. Snow is water in a solid state. Changes in temperature cause the snow to melt into liquid water. The water runs down from the mountains.

- Water in its liquid state flows down into rivers and lakes. The rivers take the water down to the sea. Some water seeps down under the ground. Plants use this water, or it is brought back to the surface of the Earth through wells and boreholes.
- The heat from the Sun causes water on the surface of the sea (and lakes and reservoirs) to evaporate and change into water vapour. Most of the water vapour comes from the sea because it has such a large surface.
- Water vapour rises up in the sky and reaches cooler air. The cooler air temperature causes the water vapour to condense and form droplets of water. These droplets of water move together and get larger and heavier.
- The water droplets fall as rain, or they can freeze and fall as snow. Cold temperatures cause the water to change into this solid state.
- If there is time, let students work in groups to make a collage of a water cycle. They can cut out or draw pictures, or use different fabrics to show the sea, rivers, snow, clouds and so on. They could label their collages and add arrows to show the cycle.

Graded activities

■1 The students should work in pairs to explain the water cycle, with the aid of a diagram. They should be able to identify the processes of condensation and evaporation in this cycle and to name the states of water in each part of the cycle.

●2 The students should work in groups and build a model to illustrate the water cycle. They should follow the instructions in their Workbook on page 36. Walk around, assisting as necessary. Ask questions about what the model shows. Ask: *Where do evaporation and condensation occur? Why? What does the plant do? Why did we put soil in the bottle?*

▲3 Ask the students to write a story, imagining that they are drops of water. The story should tell the journey of a drop of water in the water cycle. Students can make their stories into little booklets and illustrate them. These can be circulated in class for others to read and peer review.

Consolidate and review

- Students can use page 37 in their Workbooks to revise and consolidate what they have learned about the water cycle.
- Let students complete DVD Activity C3 about the water cycle to consolidate what they have learned.

Differentiation

■ All of the students should be able to explain that water moves in a cycle and that the processes of evaporation and condensation are essential parts of this cycle. Students should be able to interpret a water cycle diagram, and draw and label their own basic diagram to illustrate the process.

● Most of the students should be able to work in groups, following instructions, to make a model to illustrate the water cycle. They should be able to relate the model to the real water cycle on Earth.

▲ Some of the students should be able to write imaginative stories based on the water cycle, bringing real elements of this cycle into their stories.

2.6 Boiling and freezing

Student's Book pages 40–41

Chemistry learning objective

- Know that the boiling point of water is 100 °C and the melting point of ice is 0 °C.

Resources

- Workbook pages 38 and 39–40
- DVD Activity C4: The Celsius scale

Classroom equipment

- thermometers
- accurate thermometer or temperature sensor
- beaker of water with heating apparatus and support
- glasses of ice
- for the investigation, each group will need a pot of ice, a thermometer, a heater (gas burner, candle, stove) and support, and a stopclock or watch with a timer
- eye protection

Scientific enquiry skills

- *Obtain and present evidence:* Measure volume, temperature, time, length and force; discuss the need for repeated observations and measurements; present results in bar charts and line graphs.
- *Consider evidence and approach:* Begin to evaluate repeated results; recognise and make predictions from patterns in data and suggest explanation using scientific knowledge and understanding.

Key words

- **temperature**
- **degrees**
- **thermometer**
- **Celsius**
- **boiling point**
- **melting point**
- **freezing point**

Warn the students that spillages will make the floor slippery. Make sure all spillages are cleaned up promptly. Make sure students are aware of the dangers of using heat sources, and that they treat all hot surfaces with caution. Supervise the groups carefully, ensuring that they wear eye protection and stand up to carry out experiments.

Scientific background

Boiling and freezing will change the state of water. Pure water freezes at 0 °C to form ice (a solid) and boils at 100 °C to form water vapour (a gas). This at a generalisation that is appropriate for students in Stage 5, but it is only partly true. The actual *melting/freezing point* and *boiling point* are dependent on both *temperature* and pressure. Some children may know that at high altitudes (for example, on the tops of mountains) water boils at a much lower temperature.

The Cambridge Primary Science curriculum framework uses the *Celsius* scale (°C) for temperature, but temperatures can also be measured using the Fahrenheit (°F) or Kelvin (K) scales.

	Celsius	Fahrenheit	Kelvin
Boiling point	100 °C	212 °F	373.15 K
Melting point	0 °C	32 °F	273.15 K

Introduction

- Explain to the students that they are going to investigate the temperatures at which water changes state.
- Talk about which temperatures could be considered to be very hot or very cold. Ask: *Have you ever said 'It's freezing today' or 'It's boiling outside' or 'I'm melting in here?'* What did they mean by this?
- If you are near the sea, a river or lake, ask: *What temperature do you think the water is? Is it warm or cold?* Ask: *Would you prefer to swim or bathe in water that is 10 °C or 25 °C? Why?* Help students to understand that 25 °C is a comfortable temperature and that 10 °C is quite cold.

Teaching and learning activities

- Before you teach the students about boiling and melting points, revise temperature and the way in which we measure temperature. Ask: *What is the temperature today? How do you know?*
- Let students read temperatures on *thermometers*. Let them measure the temperatures inside the classroom and outside, and compare their readings. Remind them about the need to always take accurate readings.
- Check that students know the words *degree* and 'Celsius' and the abbreviation °C.
- Teach the students about the boiling point of water. Ask: *What happens to water when it boils? Does water evaporate more quickly when water boils? Why?*
- You could do a demonstration to measure the temperature of boiling water. Make sure the students stand at a safe distance while you do this. Use a sensor or an accurate thermometer and take a few readings to demonstrate the need for repeated readings to get accurate results. Let the water boil for a few minutes and then take another reading to demonstrate that the temperature remains the same once the water is boiling (around 100 °C).
- Then teach the students about the melting point of ice. Relate this back to the water cycle. Ask: *What causes snow or ice to melt?*

Graded activities

■ 1 The students should work in pairs and explain the difference between boiling point and melting point to their partner. Afterwards, let them write their own explanations in their exercise books. Check these explanations to see if they are accurate.

● 2 The students should place a glass of ice in the classroom. They measure the temperature inside the glass and the air temperature inside the classroom and predict what will happen to the ice. Then let them complete the activity on page 38 of their Workbooks. This requires them to read the results of an investigation and to interpret them.

▲ 3 The students should work in groups to set up an investigation to measure the relationship between temperature and the state of water. Students can set up the investigation and record their measurements in their Workbooks on pages 39–40. They should heat the water *very slowly* in order to measure the changes, stopping at the point where all the ice has melted. Afterwards, discuss the need for repeated observations and measurements. Let groups compare their results. If the results are different, discuss why this may be so. Suggest that the groups repeat the investigation if it seems that the results were not recorded accurately.

NOTE: You could take this investigation further by letting the water reach boiling point. This is best done as a demonstration as it would be dangerous for the students to do this themselves. Measure the temperature of the water at regular intervals until it boils and write the temperatures on the board.

Consolidate and review

- Students can each write down a few questions about what they have learned about boiling and melting points and carry out a short quiz in groups. Check the questions and the answers to make sure the students have understood the concepts.
- Let the students complete DVD Activity C4 to consolidate what they have learned.

Differentiation

■ All of the students should be able to explain the difference between boiling point and melting point, and to state the boiling and melting points of water.

● Most of the students should be able to set up and conduct a simple investigation to measure and record the changes of temperature as water changes state. They will be able to interpret the graph of results in their Workbook with some help.

▲ Some of the students should be able to investigate the way in which the temperature of ice changes when it is heated with little or no help. They will record their results in their Workbook.

2.7 What happens to substances dissolved in water?

Student's Book pages 42–43

Chemistry learning objective

- Know that when a liquid evaporates from a solution the solid is left behind.

Resources

- Workbook page 41
- Slideshow C1: Making salt

Classroom equipment

- transparent containers
- water
- spoons/sticks for stirring
- solids like sand, sugar, soap powder, clay, pepper, salt

Scientific enquiry skills

- *Plan investigative work:* Make predictions of what will happen based on scientific knowledge and understanding, and suggest and communicate how to test these; collect sufficient evidence to test an idea.
- *Consider evidence and approach:* Decide whether results support predictions.

Key words

- **solution**
- **dissolve**
- **solute**
- **solvent**

⚠ Warn the students that spillages will make the floor slippery. Make sure all spillages are cleaned up promptly. Students should not taste any of the solutions or mixtures. Remind them that tasting liquids in the laboratory can be dangerous.

Scientific background

When a solid is added to water, the water particles surround the solid particles at the edges of the solid structure. If the forces of attraction between the water particles and the solid particles are greater than the forces between the solid particles themselves, then the solid will *dissolve*, forming a *solution*. The dissolved solid is called the *solute* and the water is the *solvent*. There is always a limit to how much of a particular solid can be dissolved in a fixed amount of water. When no more solid will dissolve, the solution is said to be saturated. When the water evaporates, the solids dissolved in the water are left behind.

Seawater contains large amounts of sodium chloride (salt). When the water evaporates, salt remains behind. The largest sea salt flats are in Bolivia in South America. Salt is also mined in some countries.

Introduction

- Explain to the students that they are going to investigate what happens to solids when they are mixed with water. Students should be familiar with the concepts in this unit, so this is mainly revision.
- Ask: *What does the word 'dissolve' mean? What substances dissolve in water? How do we know something has dissolved?*
- Talk about the sugar that people put in tea or coffee to make it sweet. Ask: *What happens to the sugar?* Encourage the students not to say that the sugar has 'disappeared', because the sugar is still there – you just cannot see it. If any students question this, you could point out that the sugar cannot have disappeared, because we can still taste it.

Teaching and learning activities

- Use the pictures on Student's Book page 42 and, if possible, demonstrate each step in the process. Make sure the students understand the term 'dissolve'. Ask them to predict whether each substance will dissolve before you demonstrate.

(Don't use sugar as the students will use that in their own investigation.) Compare the end products to show that if a substance dissolves, the result is a clear liquid (even if it is coloured), but a substance that does not dissolve can still be seen in the water.

- Once you have revised the concepts of solutions and dissolving, move on to talk about what happens to substances that are dissolved in water when the water evaporates. Ask students to say what they think happens. Ask: *Where does the salt that we put on our food come from? How do we get this salt?* Help them to understand that the solid salt we use is what remains behind when salty water evaporates.
- Discuss the photograph in the Student's Book. Explain to the students that sea water contains large quantities of dissolved salt. One of the easiest ways of getting salt is by evaporating the water. If the water evaporates quickly, salt crystals are formed. These can be collected, and this is what we use on our food. (In some countries, salt is also mined as it is found under the ground.)
- Show the Slideshow C1: Making salt. Ask the students to identify the processes they can see. Ask: *Which processes are reversible?*

Graded activities

1 Students work in groups and set up their own investigations to test what happens when sugar is mixed with water. This is a very simple investigation that the students should be able to conduct by themselves. They record their findings in their Workbooks on page 41.

2 Students work in pairs and discuss how salt is 'farmed', as shown in the picture in their Student's Books.

3 Students can do their own research into salt production in different places. If appropriate, help them to find out about salt production methods in your own country, and how and where these are carried out.

Consolidate and review

- Ask the students to produce an independent piece of writing on salt. They can write a factual information sheet or a story.

Differentiation

■ All of the students should be able to describe what happens when we mix solids with water, to give examples of solids that dissolve in water and to say what happens to the solids dissolved in water when the water evaporates. They should be able to set up a simple investigation in groups to see if sugar dissolves in water.

● Most of the students should be able to describe in a simple way how salt is 'farmed' and mention the role of evaporation in this process.

▲ Some of the students should be able to conduct their own research into salt production and explain in more detail how the process works.

Big Cat

The students who have read *Big Cat The Sneezles* should recognise that the Royal Wash-Man uses soap and water to wash the clothes. Water is being used a solvent.

2.8 Getting the solid out of a solution

Student's Book pages 44–45

Chemistry learning objective

- Know that when a liquid evaporates from a solution the solid is left behind.

Resources

- Workbook page 42
- PCM C1: Making sugar crystal candy
- PCM C2: Making salt crystals
- Slideshow C2: Crystals
- Slideshow C3: Sugar crystal candy
- DVD Activity C5: Solutions

Classroom equipment

- salt and sugar
- clean jars
- water
- spoons/sticks for stirring
- heat source
- pans or jugs
- food colouring (optional)
- cotton or woollen string (*not* nylon)
- pencils
- eye protection

Scientific enquiry skills

- *Plan investigative work:* Make predictions of what will happen based on scientific knowledge and understanding, and suggest and communicate how to test these; use knowledge and understanding to plan how to carry out a fair test; identify factors that need to be taken into account in different contexts.
- *Consider evidence and approach:* Interpret data and think about whether it is sufficient to draw conclusions.

Key words

- **solvent**
- **solute**
- **saturated solution**
- **crystal**

Warn the students that spillages will make the floor slippery. Make sure all spillages are cleaned up promptly. Students should not taste any of the solutions or mixtures. Remind them that tasting substances in the laboratory can be dangerous. Make sure students are aware of the dangers of using heat sources, and that they treat all hot surfaces with caution. Supervise the groups carefully, ensuring that they wear eye protection and stand up to carry out experiments.

Scientific background

Crystals are formed as *saturated* or supersaturated *solutions* are cooled. These solutions are made by heating a *solvent* and dissolving as much *solute* in it as possible. Cooling the solution forces the solute out of solution in the form of crystals. The rate of evaporation affects the formation of crystals. The faster the solution is cooled, the quicker the crystals will form, but they will tend to be smaller.

Saturated solutions are covered in this unit for enrichment and for additional information only. Students do not need to know about saturated solutions or about the concept of temperature and rate of evaporation affecting the size of crystals produced as part of the Cambridge Primary Science curriculum, and will not be tested in the Progression or Primary Checkpoint tests.

Introduction

- Ask the students to discuss in pairs the meanings of the following terms: 'solute', 'solvent', 'solution' and 'saturated solution'. Take feedback from individual students and write the meanings of the words on the board.

Teaching and learning activities

- Let students discuss what they would do to make a saturated solution. If there is time and you feel that the students are unsure about saturated solutions, you could demonstrate this in class using salt or sugar.
- Let students look at the pictures of crystals in their Student's Book. Ask students to describe the crystals (colour, shape, texture, surfaces). Ask: *How do we get crystals? What affects the crystals as they form?*

- Show Slideshow C2 of some different crystals for the students to look at and comment on.
- Then look at the pictures in the Student's Book which show students investigating how to make crystals. Ask: *How have the students set up their investigation? Are they working in a safe way? What factors are they investigating?*
- Explain that temperature, the rate of evaporation and the solute substance may be factors that could be investigated, as these may all affect crystal formation. Discuss ways in which these factors might be changed, for example establish that to change the temperature the solution could heated, or cooled using an ice bath, or placed in a fridge. The rate of evaporation could be increased using containers with different surface areas, or by using a fan or hair dryer.

Graded activities

As these activities are practical investigations, it is recommended that the students work in mixed ability groups. Differentiation should be via the level of support the students receive as they work on the activity, as well as by outcome (please see guidance in the 'Differentiation' box right).

■ 1 Give each group of students a copy of PCM C1 which explains how to make sugar crystal candy. Make sure the students understand what they need to do before they begin to set up the experiment. You can show them Slideshow C3 which shows some examples of some ready-made crystal candy sticks.

● 2 Give each group of students a copy of PCM C2 which explains how to make salt crystals. Make sure the students understand what they need to do before they begin to set up the experiment.

▲ 3 Let the students work in groups and plan their own investigation into making crystals, and the factors that affect the making of crystals. Tell the students that they should investigate the effect of temperature on crystal formation. Check that they have planned their investigation carefully, and that they have taken adequate safety precautions. Make sure they know how to make a saturated solution and then let them conduct their investigations. They can record their observations in their Workbook on page 42. Some more able students can investigate how to make other types of crystals, or they can investigate changing the temperature rapidly to make different sizes of salt crystals.

Consolidate and review

- Ask the students to work in pairs and make up their own true or false questions relating to crystal formation and their investigations. The pairs should take turns to ask another pair their questions.
- Let the students complete DVD Activity C5 to consolidate what they have learned about solutions.

Differentiation

■ All of the students should be able to describe what a saturated solution is and the way it is made.

● Most of the students should be able to make decisions about the equipment they need to carry out an investigation and how they can change the variables (temperature). They will work cooperatively with other members of their group and use equipment safely. Some of the students should be able to explain how different factors might affect the size of crystals formed.

▲ Some of the students will be able to work independently and help to direct their group in working as a team. They can identify variables to change, measure and control, and use equipment with precision and care. They will also be able to conduct further investigations and make other types of crystals.

Consolidation

Student's Book page 46

Chemistry learning objectives

- Know that evaporation occurs when a liquid turns into a gas.
- Know that condensation occurs when a gas turns into a liquid and that it is the reverse of evaporation.
- Know that air contains water vapour and when this meets a cold surface it may condense.
- Know that the boiling point of water is 100 °C and the melting point of ice is 0 °C.
- Know that when a liquid evaporates from a solution the solid is left behind.

Resources

- Assessment Sheets C1 and C2

Classroom equipment

- coloured pens or pencils

Looking back

- Use the Student's Book summary points to review the key things that the students have learned in this topic. Ask questions to stimulate discussion and enquiry.
- Ask students to write down five questions about things they have learned in this topic. Let them exchange questions and answer each other's. They should check the answers themselves.

How well do you remember?

You may use the revision and consolidation activities on Student's Book page 46 either as a test, or as a paired class activity. If you are using the activities as a test, have the students work on their own to complete the tasks in writing. Collect and mark the work. If you are using them as a class activity, you may prefer to let the students do the tasks orally, in pairs. Circulate as they discuss and answer the questions, observing the students carefully to see who is confident and who is unsure of the concepts.

Some suggested answers

1 Iceberg: solid state, could change to liquid if the temperature increased to melting point. Puddle: liquid state, could change to gas (water vapour) through evaporation.
2 Water changes from liquid to gas during evaporation and from gas to liquid during condensation.
3 The boiling point of water is 100 °C and the melting point of ice is 0 °C. Water changes from a liquid state to a gas state at boiling point, and from a solid state to a liquid state at melting point.
4 We can extract salt from seawater and use the salt in our food.

Assessment

A more formal assessment of the students' understanding of the topic can be undertaken using Assessment Sheets C1 and C2. These can be completed in class or as a homework task.

Students following Cambridge International Examinations Primary Science Curriculum Framework will write progression tests set and supplied by Cambridge at this level, and feedback will be given regarding their achievement levels.

Assessment Sheet answers

Sheet C1

1 true / false / true [3]
2 gases / temperature / surface / air [4]
3 Water changes from liquid to gas during evaporation [1] and from gas to liquid during condensation. [1]
4 true / false [1]
5 0 °C [1]
6 Check students' diagrams. [4]

Sheet C2

1 false / false / false / true [4]
2 When water dries up, we say it has evaporated. Factors such as temperature, air flow and surface area affect the speed at which evaporation takes place. When a gas turns into a liquid, we say it condenses. Condensation is the reverse of evaporation. [4]
3 solutions / solute / solvent [3]
4 temperature, air flow, surface area [3]
5 gases [1]

Student's Book answers

Pages 30–31

1 Students' own answers.

2 Students' own answers.

Pages 32–33

1 The puddle will dry up or get smaller.

2 The water evaporates into the air.

3 The boys are drawing a chalk line around the puddle. They will come back later and see if the outline (the area) of the puddle has changed.

4 Temperature, air flow, surface area.

Pages 34–35

1 Drops of water form on the tile/mirror/window. Water is changing from a gas to a liquid because it is coming into contact with a cool surface.

2 Little drops of water formed on the inside of the top cup. This happened because some of the warm water in the bottom cup evaporated and turned into water vapour. The water vapour moved up into the top cup where it was cooler and the water condensed into little drops of water. If we put ice cubes on top of the top cup, condensation would be quicker and we would see bigger drops of water on the inside of the top cup. Condensation occurs when air is cooled.

Pages 36–37

1 The water changes into water vapour and mixes with the air.

2 Water evaporates from the surface of the lake and forms rain drops up in the clouds. It is falling back to Earth as rain.

3 Snow is water in a solid state. Snow is formed when water droplets that have formed in the clouds (through condensation) freeze. Snow also melts and turns into water.

Pages 38–39

1 Students should describe each stage in the water cycle: rain and snow falling; water collecting in lakes, rivers and under the ground; water running into the sea; evaporation of water from sea and lakes; condensation in the clouds.

Pages 40–41

1 Answers will vary.

2 Water evaporates more quickly at higher temperatures.

Pages 42–43

1 Sugar, soap powder and salt will dissolve. Pepper and clay will not dissolve.

2 The solute has dissolved in the water to make a solution. We can't see the solute because it has dissolved.

Pages 44–45

1 Continue to add salt to the solution until no more salt will dissolve. Heat the solution to make more salt dissolve.

2 There needs to be a lot of salt to make crystals.

3 The students first make a saturated solution. Then they warm the solution to make some of the water evaporate. Crystals start to form on the string hanging in the solution.

3.1 Making shadows

Student's Book pages 48–49

Physics learning objective

- Observe that shadows are formed when light travelling from a source is blocked.

Resources

- Workbook pages 43–44
- Slideshow P1: Light sources
- DVD Activity P1: Sources of light

Classroom equipment

- blinds or curtains to darken the classroom, if possible
- bright, narrow-beam flashlight
- flashlights for the students to use
- cardboard and scissors, or modelling clay, for making objects to cast shadows
- rulers
- sticks or chalk to record the shapes of shadows outside

Scientific enquiry skills

- *Ideas and evidence:* Use observation and measurement to test predictions and make links.
- *Plan investigative work:* Identify factors that need to be taken into account in different contexts.
- *Obtain and present evidence:* Make relevant observations.
- *Consider evidence and approach:* Recognise and make predictions from patterns in data and suggest explanations using scientific knowledge and understanding.

Key words

- **light**
- **shadow**
- **source**
- **beam**

⚠ Students should always take care not to shine light directly in each other's eyes and never to look directly at the Sun or any other light source. If the students go outdoors to observe their own shadows, ensure they are safe and that they stay together. Warn students that light sources get hot and could burn them if touched.

Scientific background

Light is a form of electromagnetic energy that comes from a light *source*. We cannot see things if there is no light source. We call this darkness – darkness is the absence of light. Our ability to see things is due to the light rays that objects reflect into our eyes. Students often misunderstand vision, and may think that light comes from our eyes.

Light travels in straight lines. It passes through some materials (these are transparent), is partially stopped by translucent materials and is totally blocked by opaque materials, creating *shadows*.

Shadows are rarely completely dark, even when the main light source is blocked, because usually there is some light reflecting into the shaded area from elsewhere. Shadows get bigger as an object is moved closer to the light source. Light spreads out from a light source, so a near object blocks out a greater proportion of the light and makes a bigger shadow.

The position and length of a shadow outdoors change during the day, as the position and angle of the Sun in the sky change. In the middle of the day, the Sun is highest in the sky and so the shadow is shorter. The main focus here is on the changing position of the shadow, rather than on its size.

Introduction

- Start by finding out what the students already know about light and dark, light sources and shadows. Ask: *When can you see things well?* (When there is light.) *When can't you see well?* (When there is no light.) Darken the classroom. Ask: *Can you see well now?* (No.) *Why not?* (There is no light.) Then use a lamp or flashlight and create shadows on the wall of the classroom by putting your hands or an object in the *beam* of light. Ask: *What can you see now?* (Shadows.) *Why?* Let students make suggestions as to the process by which the shadows were formed. The light shining on the object causes the shadow because the object blocks the light.

- Show Slideshow P1 about light sources, and discuss the pictures with the class.
- Let the students complete DVD Activity P1 to remind them about different sources of light.
- To demonstrate the need for a light source to see things, shine a bright, narrow-beam flashlight around the room (upwards, away from eyes). This also illustrates light travelling in straight lines.

Teaching and learning activities

- Ask students to look at the photographs of shadows in their Student's Book. They should look at each photograph carefully and use the questions in their book to guide their discussions. Make sure that all the students understand that shadows are formed when light shines on something that it cannot pass through.

Graded activities

For these activities, it is recommended that the students work in mixed ability groups. Differentiation should be via the level of support the students receive as they work on the activity, as well as by outcome (please see guidance in the 'Differentiation' box right).

■1 The students should work in pairs. The classroom will need to be darkened for this work and the students will need flashlights. Walk around, encouraging the students to experiment by shining their flashlights on different objects and from different distances and angles. They will draw three of the shadows they make, so they will need to observe them very carefully. Some of the students could also measure the distance between the flashlight and the object, to see if this makes a difference. Students record their observations in their Workbook on page 43.

●2 Let students work in pairs and observe their own shadows outside. They may have done this in earlier stages, but it is a good way to revise what they already know. For best results, choose a sunny day for this investigation. Take the students outside and ask: *What is the light source?* (The Sun.) *Do you make a bigger shadow if you turn your side or your back to the Sun?* (Back.) *Why?* Students should do one observation early in the morning and the other observation closer to midday. They should predict that the shadows will be different, based on their prior knowledge of shadows and perhaps their knowledge that the position of the Sun in the sky changes (because of the movement of the Earth). The observations should be at least two hours apart for students to notice the differences. Remind them that data needs to be accurate. Students can record their observations and measurements in their Workbook on page 44.

▲3 Students use information they have collected from their observations and measurements to make links and suggest explanations about why shadows can be different sizes and shapes. They should realise that the shape of the object which blocks the light determines the shape of its shadow. They should be able to start understanding that shadows get bigger as an object is moved closer to the light source. The object blocks out more light and, therefore, makes a bigger shadow.

Consolidate and review

Ask students to write a paragraph about shadows to explain the way that they are formed. Students could also draw a simple diagram to show this.

Differentiation

■ All of the students should be able to describe how and why shadows are formed, using the terms 'light source' and 'block'. They will be able to do the investigation without help and record their results.

● Most of the students should be able to conduct the investigation to record their own shadows and explain why the shadows are different at different times of day.

▲ Some of the students should be able to explain why the size of a shadow can change, using their observations and the measurements they have taken.

Students who have read *Big Cat Chewy Hughie* should recall that when the people are standing outside they cast a shadow. The light from the Sun cannot pass through them.

3.2 Shadows outside

Student's Book pages 50–51

Physics learning objective

- Observe that shadows change in length and position throughout the day.

Resources

- Workbook pages 45 and 46–47

Classroom equipment

- large sheets of paper
- toy figure (or stick/pencil positioned vertically in modelling clay)
- flashlights
- rulers
- protractors
- globe and/or balls to represent the Sun and the Earth

Scientific enquiry skills

- *Ideas and evidence:* Use observation and measurement to test predictions and make links.
- *Plan investigative work:* Identify factors that need to be taken into account in different contexts.
- *Obtain and present evidence:* Make relevant observations; measure volume, temperature, time, length and force; discuss the need for repeated observations and measurements; present results in bar charts and line graphs.
- *Consider evidence and approach:* Recognise and make predictions from patterns in data and suggest explanations using scientific knowledge and understanding.

Ensure that students do not shine light directly into each other's eyes. If the students go outdoors to observe shadows, ensure they are safe and that they stay together.

Scientific background

The position of a shadow changes as the position of the Sun in the sky changes during the day. The length of the shadow changes as the angle of the Sun in the sky changes. In the middle of the day, the Sun is highest in the sky and so the shadow at this time will be shorter than at other times.

This lesson reinforces what the students have learned in the past about the changes in the position of the Sun relative to the Earth and the ways that this affects shadows. The students should understand from work in earlier stages that it is the Earth spinning rather than the Sun moving across the sky that produces the apparent movement. As the Earth spins, the Sun appears to follow an arc across the sky from the east (where it rises), to its highest point, nearly overhead, and down again to the west, until it disappears below the horizon at sunset. The arc that the Sun appears to follow is tilted to the south in the northern hemisphere and to the north in the southern hemisphere.

Introduction

- Remind the students that the size and position of shadows in sunshine are different at different times of the day. Ask students to think back to the previous lesson, when they recorded their partners' shadows outside. Ask: *In what ways did the shadows change? Why?*

Teaching and learning activities

- Do a demonstration using a toy or other small object, such as a pencil, and a flashlight to represent the Sun. Explain to the students that you are using a model to investigate shadows. Ask: *What path does the Sun follow in the sky throughout the day, due to the spinning of the Earth?* If they don't remember, mimic the Sun rising in the east and setting in the west, and travelling in an arc tilted to the south (if you are in the northern hemisphere). Point out the way that the shadow of the toy changes with the position of the flashlight.
- Let students work in pairs to discuss the pictures and the questions in their Student's Book, before they complete the graded activities.

Graded activities

1 Students set up and conduct their investigations in groups. They place their figure on to a large sheet of paper and shine the flashlight at it to create a shadow. They move the flashlight in an arc as the Sun would travel – from sunrise, to midday, to sunset – and watch what happens to the shadow. Check that the students are making a suitable arc with the flashlight. When they have practised, they repeat the process, stopping in various positions to draw round the shadow on the piece of paper. Ideally, they will draw the shadow in at least five different positions.

They could write N, S, E and W on their paper if they are able, and label the drawn positions with the approximate time of day, for example 'early morning'. Encourage students to check their observations by repeating them. Ask the students to present their findings. They should have noted that the shadow's length and the direction (angle) in which it pointed both changed.

The students record their investigations in their Workbooks on page 45. If there is time, the activity should be repeated so that each member of the group has a turn at 'being the Sun' and at recording the appearance of the shadow, and so that each has their own record of the activity. Some students will be able to measure the length of the shadow and its angle from south, at each position (or at least early in the morning, late morning and at midday), presenting their measurements in a table.

2 The students should conduct an investigation outside to practise taking accurate measurements and presenting these as charts. They follow the guidelines in their Workbooks on pages 46–47. The focus is on observing and measuring the difference in length (not direction) of the shadows and showing this on a bar chart.

3 Some students should be able to give an oral explanation of the way that the movement of the Earth makes shadows change. They can use a globe and/or some balls to demonstrate the way that the position of the Sun changes relative to the Earth.

Consolidate and review

Let the students, in their groups, present the results of their investigation to the rest of the class. They should take questions and consider any suggestions of ways to improve upon their investigations.

Differentiation

■ All of the students should be able to state that shadows are longest early and late in the day and shortest during the middle of the day, and that the angle and direction of the shadows change during the day as a result of the position of the Sun in the sky.

● Most of the students should be able to describe the way that the direction in which a shadow points changes throughout the day. They will challenge each other's ideas and come to an agreement as they conduct their investigation. They should choose to check observations by repeating the activity. They should be able to conduct their investigations with a little assistance.

▲ Some of the students should be able to describe the way that the movement of the Earth accounts for the differences in the shadows. Some of the students will rightly criticise the model, because the Sun (flashlight) is moving when, in fact, the Earth should be moving. They could demonstrate this.

3.3 Changing the size of a shadow

Student's Book pages 52–53

Physics learning objective

- Investigate how the size of a shadow is affected by the position of the object.

Resources

- Workbook pages 48–49

Classroom equipment

- flashlights or lamps as light sources
- square pieces of cardboard (5 × 5 cm)
- long pencils or sticks
- sticky tape
- modelling clay
- metre rulers
- 1 cm squared paper

Scientific enquiry skills

- *Ideas and evidence:* Use observation and measurement to test predictions and make links.
- *Plan investigative work:* Identify factors that need to be taken into account in different contexts.
- *Obtain and present evidence:* Make relevant observations; measure volume, temperature, time, length and force; discuss the need for repeated observations and measurements.
- *Consider evidence and approach:* Decide whether results support predictions; recognise and make predictions from patterns in data and suggest explanations using scientific knowledge and understanding.

Key word

- **light source**
- **cast**

⚠ Students should always take care not to shine light directly in each other's eyes and never to look directly at the Sun or any other light source. Warn students that light sources get hot and could burn if touched.

Scientific background

Shadows are bigger when the object is closer to the light source. Light spreads out from the light source, so a near object blocks out a greater proportion of the light and *casts* a bigger shadow.

Introduction

- Tell students that they are going to plan and complete a detailed investigation about changing the size of shadows. They will investigate the effect of the distance from a *light source* to a piece of card on the size of its shadow.

Teaching and learning activities

- Ask the students to look at page 52 of the Student's Book. Let them work in pairs to discuss the questions to make sure they understand what is being investigated and the way that the investigation has been set up. They should also make predictions.
- Then, demonstrate the arrangement of the puppet-show shadow investigation, as shown in the Student's Book but using a square of card on a stick, in place of the puppet. Using a square card will enable students to take accurate measurements when they do their own investigations.
- Take feedback on the students' predictions of the size of the shadow as the card is moved further from the lamp, and ask for reasons. Do not demonstrate what happens – this will be left for the students to investigate themselves.

Graded activities

For these activities, the students should work in mixed ability groups. Differentiation should be via the level of support the students receive as they work on the activity, as well as by outcome (please see guidance in the 'Differentiation' box right).

1 The students collect the materials and set up their investigation in groups. They record their predictions in their Workbook on page 48 and should use the questions there for general guidance on the investigation.

2 The students work in groups to test their predictions.

▲3 The students measure the distance between the card and the light and then note the height of the shadow. This is most easily done by casting the shadow on to a screen made from squared paper: 1 cm squares will allow the students to measure the shadows quickly and easily. They should aim to get results for four or five different card positions.

Students record their observations, measurements and conclusion on pages 48 and 49 of the Workbook, working individually.

Consolidate and review

- Once the groups have completed their investigations, ask them to discuss their results for a few minutes and then take feedback. Did all groups get the same results? The students should find that the shadows were bigger when the card was closer to the light source. Were there any difficulties with the investigation?
- Ask students to make a statement about their investigation. Some of them can do this orally and others could write about it. They should use the information recorded in their Workbook as a basis for the statement. The statement should include what they investigated, what they predicted, how they set about doing the investigation, what they recorded, and a conclusion.

Differentiation

■ All of the students should be able to take part in an investigation and take some measurements, with help. They should be able to see a trend in their results and write a conclusion, with help.

● Most of the students should be able to take care to make accurate measurements to at least the nearest half centimetre. They may discuss the best format for recording results. They may also discuss alternative predictions and conclusions, and come to an agreement. Most of the students should write a correct conclusion, working independently.

▲ Some of the students should be able to make measurements to the nearest millimetre. They may notice any difficulties or surprises that arise during the investigation, e.g. the effect of light from other groups, and may take steps to reduce the problem. They will write a conclusion and make a good attempt to explain that, as the light is spread out from the light source, a close object makes a bigger shadow than a far one.

3.4 Recording shadows

Student's Book pages 54–55

Physics learning objective

- Observe that shadows change in length and position throughout the day.

Resources

- Workbook pages 50–51
- Slideshow P2: Sundials
- PCM P1: Making a sundial
- PCM P2: Sundial template

Classroom equipment

- sticks
- real sundial, if available
- large sheets of paper to make posters
- coloured pens or pencils
- thick card
- scissors
- sticky tape or glue

Scientific enquiry skills

- *Ideas and evidence:* Use observation and measurement to test predictions and make links.
- *Obtain and present evidence:* Make relevant observations; measure volume, temperature, time, length and force; discuss the need for repeated observations and measurements; present results in bar charts and line graphs.

Key word

- **sundial**

Students should always take care to never to look directly at the Sun or any other light source. If the students go outdoors to observe shadows, ensure they are safe and that they stay together. Ensure that they take care with the sticks, and that they use scissors safely.

Scientific background

Because the Earth moves relative to the Sun in a predictable manner, shadows change predictably throughout the day. Ancient civilisations understood this and created *sundials*. Sundials can be sophisticated, and some designs can do more than just tell the time. A sundial works by using the shadow cast by an object called a 'gnomon'. The gnomon can be a thin rod or a triangular shape. The angle at which the rod is positioned needs to be adjusted to take into account the latitude of the location of the sundial. The gnomon should be pointed towards the celestial North Pole.

Introduction

- In this unit students find out more about the way that shadows can be used to tell the time. Show Slideshow P2 about sundials and show the students a real sundial if you have one. Explain that sundials only show the time during daylight hours as there are no shadows at night, and that there are many different types of sundials.

Teaching and learning activities

- Demonstrate to the students how a sundial works. You may not be able to tell the time accurately, unless the sundial has been specifically made for the area in which you live and is aligned correctly with the North Pole, but you can demonstrate the principle by which it works.
- Explain that sundials came about because people observed that shadows made on the Earth follow a predictable pattern. This is because of the way the Earth moves around the Sun in a predictable way.
- Let students work in pairs to discuss the pictures and the questions in their Student's Book, before they complete the graded activities.
- If there is time, students can do some research about ancient sundials – where they were made and who used them.

Graded activities

1 Ask students to write a paragraph about sundials, explaining, in a basic way, how they work and why people used them in the past. They can copy out their paragraphs neatly on to some poster paper and illustrate their work for a class display.

2 The students work in groups. They discuss and then set up their investigations to measure the direction and length of shadows. They can use a stick to do this. For this investigation they will need to know the compass points and to be able to establish the directions of N, S, E and W. The students should be able to observe the length of the shadow and its direction at six or more different times of the day. They should be able to make a diagram in their Workbook on page 50 to record what happened, and some may go on to draw a line graph of their findings on page 51. Encourage students to check their observations by repeating them the following day. Then ask the students to present their findings. They should have noted that the shadow's length, and the angle or direction in which it pointed, changed throughout the day.

▲3 Let the students build their own sundials. Give them copies of PCM P1 and PCM P2 to use as a template and help them to build these. Help them to align their sundials correctly and then use them to tell the time.

Consolidate and review

- Let students work in groups to draw up lists of questions to ask each other about what they have learned on the subject of shadows. Ask them to write down the answers to their questions; check these before they ask each other the questions.

Differentiation

■ All of the students should be able to explain that a sundial casts shadows which we can read to tell the time. They can copy out their work neatly and produce an informative and interesting poster.

● Most of the students should be able to set up an investigation and draw a diagram to show the direction and length of shadows cast at different times of the day, with help.

▲ Some of the students should be able to draw a line graph of the results of their investigation, and to build their own sundials with some assistance and use them to tell the time in a general way (they do not need to be 100% accurate).

3.5 Materials and light

Student's Book pages 56–57

Physics learning objective

- Explore how opaque materials do not let light through and transparent materials let a lot of light through.

Resources

- Workbook pages 52 and 53
- Slideshow P3: Glass buildings

Classroom equipment

- curtains or blinds to darken the classroom
- light source with a strong, narrow beam, e.g. a flashlight or projector
- card (black and coloured)
- other opaque materials, e.g. pieces of wood or metal
- transparent and translucent materials (colourless and coloured), e.g. thin paper, plastic, acetate, glass tiles
- samples of opaque and coloured liquids, e.g. milk, molasses, clear apple or berry juice, diluted ink
- scissors
- straws or sticks to hold flimsy materials
- sticky tape
- flashlights for students to use
- screen or plain pale wall space
- variety of examples of food packaging, to include opaque, translucent and transparent materials

Scientific enquiry skills

- *Plan investigative work:* Use knowledge and understanding to plan how to carry out a fair test; collect sufficient evidence to test an idea; identify factors that need to be taken into account in different contexts.
- *Obtain and present evidence:* Make relevant observations; discuss the need for repeated observations and measurements.
- *Consider evidence and approach:* Decide whether results support predictions.

Key words

- **opaque**
- **transparent**
- **translucent**

Take care not to shine light directly into the eyes. Warn the students that light sources – such as the projector lamp – get hot and could burn if touched.

Scientific background

Materials are transparent, translucent or opaque, depending on how much light they allow through. *Transparent* materials let light pass through them. The light rays pass through the material without being affected by the atoms or molecules within. These materials include clear glass and acrylic sheet. They can be colourless or coloured. *Translucent* materials allow some, but not all light, to pass through. Looking through these materials, you can see a blurred image. They are used, for example, for bathroom windows to give some privacy. *Opaque* materials block the light totally. These materials create dark shadows. They include bricks, concrete and wood.

Transparent coloured materials allow light of their own colour through. Translucent coloured materials form fuzzy coloured shadows, because they let some of the coloured light through, but also block some. Opaque coloured materials do not form coloured shadows (they form black shadows), because they do not allow *any* light through.

Introduction

- Remind the students that light travels in straight lines from a light source and that some materials block light as it travels; this is when we see shadows. Introduce and explain briefly what the terms 'opaque', 'transparent' and 'translucent' mean. Then let the students complete question 1 on page 56 of the Student's Book.
- Discuss why different materials are used for different purposes. Show Slideshow P3 of glass buildings and talk about why some glass is translucent and some transparent.

Teaching and learning activities

- Let students look at the pictures and read the captions in their Student's Book. Then let them work in pairs to discuss questions 2 and 3. Take feedback and let students share and question ideas.
- Ask: *What types of shadows do coloured objects make?* Darken the room and using the bright beam from a projector (or other strong light source), form a sharp dark shadow using a piece of black card or other black opaque material. Hold up a piece of opaque coloured card and ask the students to predict what its shadow will be like. If they do not suggest that the shadow might be coloured, make that suggestion and ask if any agree. Demonstrate that the shadow is still sharp and dark but is *not* coloured. Ask: *Why there is a sharp dark shadow?* (Light travels in straight lines; the card blocks all the light travelling from the source. It is opaque. Its colour does not matter.)
- Ask the students for examples of transparent, translucent and opaque materials. Ask: *Which materials will give sharp dark shadows?* (Opaque.) *Which materials will give very little or no shadow at all?* (Transparent.) *Which materials will give coloured shadows?* (Translucent, coloured.) *Which materials will give fuzzy, pale shadows?* (Translucent.)
- Demonstrate that coloured materials can be transparent and that not all opaque materials are solid. Use opaque liquids such as milk and transparent liquids such as clear apple juice or berry juice in your demonstrations.

Graded activities

1 The students should name three objects from home (or school) that are made from opaque materials and three objects made from transparent materials. They should discuss what the materials are and what the objects are used for, linking the property to the object's purpose.

2 The students should work in groups and set up their own investigations to find out whether different materials block light or let it through. They record their predictions and their observations on page 52 of their Workbook.

3 Let the students investigate the use of different types of materials for packaging and find out why materials with different properties are used. They choose two types of packaging and write notes about them, following the guidelines in their Workbooks on page 53.

Consolidate and review

Give the students some starters and let them complete the sentences. For example: *You get a sharp dark shadow when … ; You get a fuzzy pale shadow when …; You get no shadow when … ; You get a coloured shadow when … .*

Differentiation

■ All of the students should be able to name three different opaque and three different transparent materials, and be able to say why these properties are useful to the objects that are made from them. For example, a car windscreen is transparent so you can see through it.

● Most of the students should be able to work in groups to set up an investigation to find out how much light different materials let through. They should be able to identify what they will need for the investigation, predict results and discuss whether they will need to repeat observations to get reliable results. They should be able to describe the similarities and differences that exist between shadows cast by opaque, translucent and transparent materials.

▲ Some of the students should be able to investigate packaging and explain why it is used. For example, they will be able to work out that transparent packing is used so that people can see what is inside and that opaque or translucent packaging may be used when too much light could damage a product.

Big Cat

Students who have read *Big Cat Rally Challenge* should be able to say why it is important that the visor on the crash helmet and the glass in car windows and lights are transparent.

3.6 Playing with light and materials

Student's Book pages 58–59

Physics learning objective

- Explore how opaque materials do not let light through and transparent materials let a lot of light through.

Resources

- Workbook pages 54, 55 and 56

Classroom equipment

- translucent screen for puppet show (e.g. wooden frame with tissue paper screen)
- powerful flashlight or projector lamp
- opaque card
- scissors
- straws, sticks or pencils to hold up the puppets
- sticky tape
- ruler
- coloured pens or pencils
- coloured tissue paper (optional)
- variety of opaque, transparent and translucent materials, e.g. as listed on Workbook page 55

Scientific enquiry skills

- *Plan investigative work:* Use knowledge and understanding to plan how to carry out a fair test; collect sufficient evidence to test an idea; identify factors that need to be taken into account in different contexts.
- *Obtain and present evidence:* Make relevant observations; measure volume, temperature, time, length and force; discuss the need for repeated observations and measurements.
- *Consider evidence and approach:* Decide whether results support predictions.

Key word

- **fuzzy**

⚠ Ensure that the students do not look directly at the light source. Warn the students that light sources – such as the projector lamp – get hot and could burn if touched. Ensure that the students take care with the sticks, and that they use scissors safely.

Scientific background

Transparent coloured materials allow light of their own colour through. Translucent coloured materials form *fuzzy* coloured shadows, because they let some of the coloured light through but also block some. Opaque coloured materials do not form coloured shadows (they form black shdows), because they do not allow any of the light through.

Introduction

- This lesson consolidates what the students have learned about shadows and allows them to play with materials, using their knowledge to create effects.

Teaching and learning activities

- Ask students to open their Student's Book on page 58 and look at the photograph of the shadow puppets. Let the students discuss in pairs the way that shadow puppets work. Take feedback and help students to explain how the shadow puppet theatre works.
- Darken the room and prepare to demonstrate how a shadow puppet works. See page 58 of the Student's Book for illustrations of how a shadow puppet theatre is arranged. Ask the students about the type of materials needed for the puppets and the screen. (Opaque and translucent, respectively.) Ask: *Why does the screen need to be translucent?* (So that some light can shine through it.) *What materials do we need to use for the puppets? Why?* (They need to be opaque, to make shadows.) *How can we make the puppets look bigger or smaller?* (By moving them closer to or further away from the light source.) *What must be kept the same?* (The light source.) Record the students' comments and predictions.
- Measure the distance from the puppet to the light source and record the measurement, along with the height of the puppet's shadow at this distance.

Record these dimensions in centimetres in a simple table, for example:

Distance from the puppet to the screen	Height of the shadow

- As you demonstrate, record the measurements each time you change the position of the puppet.
- Carry out the demonstration and then refer back to the predictions the students made. Relate these to your measurements and discuss.
- If there is time, allow students to use the equipment and set up their own shadow puppet show for other classes. They can use a story they know well and make puppets of the main characters. Students will need some help to set this up and it will take a lot of time. It can be linked to language, literacy and literature studies.
- Point out the pictures of the stained-glass windows in the Student's Book. Ask students if they have seen any windows like this. Ask: *Where have you seen windows like this?* (Often in places of worship and other public buildings.) *Why are the windows made like this?* (They are decorative and they allow light into the building.)
- Let students work in pairs to discuss questions 4 and 5. Take feedback and discuss the answers as a class.

Graded activities

1 The students should draw a picture of their own stained-glass window design. This activity allows students to be creative. They can draw their designs in their Workbooks on page 54. Make sure that they answer the questions on the Workbook page, as these help the students to make scientific statements about what they have drawn. If there is time, students could make their 'windows' using coloured sheets of translucent paper and use a flashlight as a light source to see what happens when light shines through them.

2 In groups, the students should conduct a further investigation into the types of shadows that can be created using different materials. They need to decide what they will investigate, collect the materials they need and record their observations in their Workbooks on page 55. To extend this activity, ask students to take measurements as well and to think of a way of recording the measurements.

3 Let the students use the equipment you used in your demonstration to set up their own shadow puppet theatre. They can then explain how it works by drawing a diagram and writing a short explanation.

Consolidate and review

- Students can use page 56 in their Workbooks to consolidate and review what they have learned about shadows and light.

Differentiation

■ All of the students should be able to design a stained-glass window on paper, describe the light source and what happens to the light when it moves through the coloured glass.

● Most of the students should be able to work in groups and set up another investigation to find out how much light different materials let through. They should be able to identify what they will need for the investigation, predict results and discuss whether they will need to repeat observations to get reliable results. They should be able to describe the similarities and differences that exist between shadows with opaque, translucent and transparent materials and explain, in a simple way, how a shadow puppet show works.

▲ Some of the students should be able to give complete explanations of shadow puppet theatres and be able to set one up themselves.

3.7 Can we measure light?

Student's Book pages 60–61

Physics learning objective

- Know that light intensity can be measured.

Resources

- Workbook pages 57 and 58

Classroom equipment

- light meters
- access to the internet or reference books

Scientific enquiry skills

- *Obtain and present evidence:* Make relevant observations; discuss the need for repeated observations and measurements.
- *Consider evidence and approach:* Begin to evaluate repeated results; recognise and make predictions from patterns in data and suggest explanations using scientific knowledge and understanding.

Key words

- **intensity**
- **light meter**
- **lux**

Students should never look directly at the Sun or any other light source. If the students go outdoors to measure light intensities, ensure they are safe and that they stay together.

Scientific background

The *lux* (symbol, lx) is the SI unit of illuminance and luminous emittance, measuring luminous flux per unit area. It is equal to one lumen per square metre. It is used to measure the *intensity* of light as seen by the human eye. In English, 'lux' is used in both singular and plural.

The candela (symbol, cd) is the SI base unit of luminous intensity, which is the power emitted by a light source in a particular direction, weighted by the luminosity function. A common candle emits light with a luminous intensity of roughly one candela. The word 'candela' means candle in Latin, as well as in many modern languages.

Introduction

- Ask students to look around and to look through the windows outside. Ask: *Is it a bright day or a dull day? Why?* Help students to decide whether the light is bright or dull. Talk about the light sources that are making the light bright or dull.
- Teach the word 'intensity', explaining that this is how bright or strong the light is.

Teaching and learning activities

- Show the students a *light meter* and explain how it works.
- Teach the students about the lux, as a unit of measurement. Explain that it measures the intensity of light that we see with our eyes. It does not measure the energy that is coming from the light source.
- Demonstrate the way to use a light meter. Take readings in different parts of the classroom and outside, and write them on the board. Help the students to understand the range of readings that is possible. They should realise that the intensity of the light outside in the bright sunlight will give a reading of 10 000 lux, or more. Even if it is overcast and rainy outside, the intensity of light will be high. The intensity of light inside a building will be much lower, usually lower than 1000 lux. If possible, take a reading inside a dark cupboard to show that the reading can drop to below 1 lux.
- Draw a table on the board, like the one shown on the next page, and help the students to predict what sort of readings one might expect at these times. These should be very general. Help students to think about the intensity of the light at each time.

Time	Predicted lux reading
In the house when you wake up in the morning	
Outside the school at break time	
At night, when there is a moon	
In a local shop, during the day	

- Students work in pairs to discuss questions 2 and 3 in their Student's Book. Ask the pairs to report back on the usefulness of light meters. They should study the pictures in their Student's Book for ideas. Those who are familiar with photography may be able to suggest why photographers use light meters. Note, though, that light meters are built into most modern cameras and that lux readings are not shown on the camera. (Students will learn more about the uses of light meters in the next unit.)

Graded activities

1 The students should arrange the lux measurements on page 61 of the Students' Book in order from the brightest to the dullest light.

2 Let the students work in pairs or small groups and measure the intensity of light at different places in the classroom (or outside). Ask students why it is a good idea to take more than one measurement in each place. Let students compare and discuss their results. Students can record their readings in their Workbooks on page 57.

3 Ask the students to do some research to find out about other units of measurement, for example the candela. Some of the information about these units is quite complex, so help them to find accessible sources of information.

Consolidate and review

- Let students take further measurements using a light meter if they are not yet confident about doing this. Ask them to predict first before they use the light meter to measure. This will help them understand what to expect and, therefore, to record measurements more accurately.
- Students can also answer the questions in their Workbook on page 58. Explain that they have to choose the best word or number to complete each sentence. Tell the students the correct answers afterwards and let them check their own work.

Differentiation

■ All of the students should know that the brightness or intensity of light can be measured, and that we usually measure this in lux. They should be able to put lux measurements in order to show that they understand the way the measurements work.

● Most of the students should be able to use a light meter to measure the intensity of light in different places. They should be able to take accurate readings and record these. They should understand the need for repeated readings and be able to compare their readings with those of other students.

▲ Some of the students should be able to research and describe other units for measuring light, like for example the candela.

3.8 When do we need to measure light intensity?

Student's Book pages 62–63

Physics learning objective

- Know that light intensity can be measured.

Resources

- Workbook page 59

Classroom equipment

- some photographs taken at different exposures
- light meter
- a few cameras for the students to use, if possible
- an older or professional manual camera to demonstrate how the amount of light allowed into the camera can be controlled

Scientific enquiry skills

- *Plan investigative work:* Identify factors that need to be taken into account in different contexts.
- *Obtain and present evidence:* Make relevant observations.
- *Consider evidence and approach:* Recognise and make predictions from patterns in data and suggest explanations using scientific knowledge and understanding.

Key words

- **light meter**
- **camera**

If the students go outdoors to take photographs, ensure they are safe and that they stay together.

Scientific background

The intensity of light, as seen by the human eye, can be measured; light measurement is used in photography, horticulture and other areas of human activity.

The students do not need to learn about the use of *light meters* in daily life as part of Cambridge Primary Science curriculum for Stage 5. This lesson goes a little beyond the requirements of the curriculum and is included here for enrichment purposes and to place the objective in a familiar context so that students can apply their learning to real-life situations.

Introduction

- If you are familiar with photography, demonstrate how light is important in photography. You could do this by taking a series of photographs of the same place, person or object at different exposures. Explain that the amount of light that was allowed into the *camera* was different for each photograph.
- Alternatively, invite a photographer to come to speak to the class and to demonstrate the importance of light in photography. The photographer can demonstrate the use of a light meter as well.

Teaching and learning activities

- Use the sample photographs you have brought and any cameras you have available to teach students more about light in photography.
- Tell the students that light meters are built into most modern cameras and explain that most cameras do all the measurement automatically, selecting the best settings for each picture. If we want to change the amount of light allowed into the camera to create a special effect, we can override the automatic settings.
- Then discuss other uses of light meters.
- Then ask the groups to contribute to a general class discussion. Ask: *Do plants make food at night?* (No.) *Why not?* (There is no light.) *Do plants grow well in winter?* (No.) *Why not?* (It's too cold and there is not enough light.) Help students to focus on the importance of sunlight for plant growth. Then explain that the intensity of light needs to be in the region of 1200–2000 lux for this to take place successfully. Ask: *In what way could you apply this scientific knowledge?* Help students to focus on commercial horticulture, in which plants are grown inside; the intensity of light inside a building can be increased to the level needed for plants to grow

successfully. (Some plants can grow at lower light intensities.)

- Let students work in pairs or groups to discuss ways in which architects could use light meters. Take feedback in a general class discussion afterwards. Ask: *What is the average lux reading inside an office? Why do we have electric lights inside buildings?* Help students to focus on the need to have the correct light intensity inside a building so that people can live and work comfortably.

Graded activities

For these activities, the students should work in mixed ability groups. Differentiation should be via the level of support the students receive as they work on the activity, as well as by outcome (please see guidance in the 'Differentiation' box right).

1 The students should complete the exercises on page 59 of their Workbooks individually to consolidate and review what they have learned.

2 Let the students explain to their group why photographers use light meters. If possible, let them use a small camera to take digital photos to illustrate their explanations.

3 Give each group of students a camera. Each member of the group should take at least one photograph in locations where the light intensity is different. For example, they could take photos in dark places without the aid of a flash to show what happens when the light is not intense enough. They could also take photos in a very bright place and look at what happens when there is too much light. If students know enough about manual settings on cameras, they can also experiment with using different apertures to control the amount of light. The students discuss the photographs, and discuss ways to achieve a desired effect by changing the amount of light.

Consolidate and review

- Write the following sentences on the board and ask the students to fill in the gaps.
  - a A light meter measures the __________ of light. It measures this in units called __________.
  - b Light intensity is important in plants. Plants need a light intensity of between __________ and __________ to make their food.

Differentiation

■ All of the students should be able to explain, in a simple way, what light meters measure. They should also be able to name at least two ways in which people use light measurements.

● Most of the students should be able to explain why photographers use light meters and explain why measuring light can be important when growing plants.

▲ Some of the students should be able to take light measurements and use these to take photographs to achieve a desired effect. They will be able to explain this to less able students and help them to achieve the desired results.

3.9 Seeing light

Student's Book pages 64–65

Physics learning objective

- Know that we see light sources because light from the source enters our eyes.

Resources

- Workbook pages 60 and 61
- Slideshow P4: How the eye works

Classroom equipment

- a candle in a candle stand or tray of sand, matches
- shoebox painted matt black inside and with a very small hole at one end, a larger hole underneath, and a small object dangling from the lid inside the box
- flashlight
- coloured pens or pencils

Scientific enquiry skills

- *Ideas and evidence:* Know that scientists have combined evidence with creative thinking to suggest new ideas and explanations for phenomena.
- *Plan investigative work:* Make predictions of what will happen based on scientific knowledge and understanding, and suggest and communicate how to test these.

Key words

- **light source**
- **bounce**
- **reflect**

Students should not touch each other's eyes or use any sharp objects near their eyes, and should not look directly at any light source.

Scientific background

The Greek school of Pythagoras (sixth to fourth century BC) made one of the first attempts to explain the behaviour of light. Their theory was that sight could be explained by the human eye sending out a beam of light to the object being viewed. This light then streamed back with information about colour and shape. We now know that this theory is incorrect.

Light travels from a source. We see objects because light from them enters our eyes and our eyes are sensitive to light. If the object is a light source, the light that enters our eyes originates from the object. For other objects, light from another source is reflected by the object and enters our eyes. This is true even of objects that do not seem to be shiny. There must be some light for us to see objects. In total darkness, we can see nothing. When we see colours, we are seeing light of different wavelengths. Light is a type of electromagnetic radiation; other types include infra-red (longer wavelength) and ultraviolet (shorter wavelength), with coloured visible light inbetween.

Light comes into our eyes, passing though the cornea, the pupil and the lens. The lens focuses the light on to the retina, which is at the back of the eye. The retina contains millions of cells, which respond to light by transmitting messages along the optic nerve to the brain. The brain interprets the messages and creates the image that we see.

Students do not need to know about the structure of the eye as part of the Cambridge Primary Science curriculum and it will not be tested in the Progression or Primary Checkpoint tests. This information is included as additional information only.

Introduction

- Light the candle and show it to the students. Ask: *Is this a light source?* (Yes.) Then blow out the candle and ask the same question. (No.) Ask students to name other light sources.
- If appropriate, or the students ask for more information you could show them Slideshow P4 about how the eye works. The students do not need to know how the eye works as part of the Cambridge Primary Science curriculum and it will not be tested in the Progression or Primary Checkpoint tests. The slideshow presentation is included here as an optional activity and is for additional information and enrichment purposes only.

Teaching and learning activities

- Draw a diagram of the candle on the board and show light going outwards from its flame (with straight arrows). Some of the light should be going towards an eye, drawn on a face near the candle. Explain the process, stressing that we see the candle because it gives out light that goes into our eyes.
- Light the candle again and blow it out again. Ask: *How can we see the candle when there is no flame?* Let the students work in pairs to discuss this. Ask them to draw a diagram to show what they think happens. Some may think that the eye emits a light of its own, others may think that the light comes from elsewhere. Encourage the students to formulate a statement describing the process by which we can see things that are not light sources.
- Change the drawing on the board to show a blown-out candle with no light going towards the eye. Draw in a source of light, such as the Sun or the classroom lights, and show a ray travelling from the source to the candle, then to the eye. Use arrows to show the direction in which the light travels. Ask the students whether they think this might be the way that we can see the candle. Tell the class that they are going to collect evidence and then evaluate the different views.
- Set up the shoebox. Let some students complete question 1 in the Student's Book while others come up in groups to look at the shoebox. They look through the small hole in the shoebox while you cover the larger hole underneath with the unlit flashlight. The eyehole should be so small that the students are unable to see anything inside. Ask: *Can you see what I have put inside the box? What do we need for you to be able to see it?* If they say light, switch on the flashlight but ask them to not to say what is in the box until everybody has had a turn.
- When everyone has had a chance to look through the hole in the shoebox, discuss what the students observed. Ask: *Could you see anything the first time you looked?* (No.) *Why not?* (There was no light source.) *Could you see anything when the flashlight was switched on?* (Yes.) *Why?* (There was a light source – the flashlight.) *How did the light reach your eyes?* (Through the small hole.) *Did the light come from your eyes?* (No, it came from the flashlight.)

  Help the students to conclude that this investigation provided evidence that we cannot see when there is no light source and, therefore, the light does not come from our eyes.
- Ask the students to study the picture of the boy reading on page 64 of the Student's Book. Let them trace the path of the light.

Graded activities

■1 The students should work in pairs. They should draw a picture of their partner's eye. Discuss with them why they think there is a hole in the front of the eye. Help them to understand that this is where light enters our eyes.

●2 Ask the students to complete the diagrams and sentences on page 60 of their Workbooks to consolidate what they have learned about how we see things.

▲3 Let the students draw pictures of two different light sources and answer the questions on page 61 of their Workbooks.

Consolidate and review

- Let the students work in pairs and explain to their partners how they are able to see certain objects inside and outside of the classroom.

Differentiation

■ All of the students should be able to draw a picture of their partner's eye. They should understand that we see because light from a source enters our eyes.

● Most of the students should be able to complete the activity in their Workbook independently.

▲ Some of the students should be able to draw two light sources and explain why they are light sources with little or no help. They will be able to confidently answer the questions with some help.

3.10 Reflecting light

Student's Book pages 66–67

Physics learning objectives

- Know that beams/rays of light can be reflected by surfaces including mirrors, and when reflected light enters our eyes we see the object.
- Know that shadows are formed when light travelling from a source is blocked.

Resources

- Workbook pages 62 and 63
- Slideshow P5: Reflected light

Classroom equipment

- curtains or blinds to darken the classroom, or access to a darkroom
- flashlight and book
- pictures or models of the Earth, the Sun and the Moon
- coloured pens or pencils
- access to a dictionary and the internet or reference books

Scientific enquiry skills

- *Obtain and present evidence:* Make relevant observations.
- *Consider evidence and approach:* Recognise and make predictions from patterns in data and suggest explanations using scientific knowledge and understanding.

Key words

- **bounce**
- **reflect**

Remind students never to look directly at the Sun.

Scientific background

When light travels from a light source and hits an object, it can either be absorbed, transmitted through it, or *reflected* (*bounced* back). Objects that reflect or absorb the light without transmitting any of it are called opaque. Those through which light is transmitted are transparent or translucent. However, all objects reflect some light. This is something that is by no means obvious and is hard for students to appreciate. However, without this reflected light, we would not see objects at all – they would be invisible.

Students may be confused between shadows and images. In the formation of a shadow, an opaque object is blocking the light from a source. In the formation of an image, the light from an object is changing direction before entering the eye.

Many students will not have experienced complete darkness. Street and car lights and well-lit houses have created a generation of people who have never experienced true darkness. If travelling in the country, there is often a glow from distant towns and cities in the night sky.

Introduction

- If you can, darken the classroom or take the students in groups to a darkroom. Be sensitive to students who may be afraid of dark places. Use a book and a flashlight. Switch off the lights, open the book on any page and ask a student to tell you what is on that page. (The student should not be able to see very much at all.) Then shine the flashlight on the page and ask the student the same question. Ask: *Why can [name] see what is on the page?* (There is light.) *Where is the light coming from? How does it get to their eyes?* Help them to understand that the light bounces off the page and to the student's eyes.

Teaching and learning activities

- Revise how light travels from a light source, reflects off a surface and enters our eyes.
- Students study the picture of the bicycle reflector and discuss how we can see it. Let students do this in pairs first. Take feedback and make sure all the students understand how we can see the reflector.

- Let students read the comic strip story in their Student's Book in pairs. Then take feedback and let students explain what they learned from the story. Do they agree with what the characters say? Why or why not?
- Then use your models or pictures of the Sun, the Moon and the Earth to explain the way that the Moon reflects light from the Sun. Ask: *Is the Moon a light source?* (No.) *Why does it shine/give us light?* (It reflects light from the Sun.) Demonstrate and explain why sometimes we can only see part of the Moon.
- At this stage, you could explain what an eclipse is, for enrichment, although it has to do with blocked light, rather than reflected light. Eclipses happen because light travels in straight lines. Sometimes the path of the light is blocked by an opaque object that stops the light of the Sun. In a solar eclipse, the Moon blocks light from the Sun and prevents it from reaching the Earth. In a lunar eclipse, the Earth blocks the Sun's rays from travelling to the Moon. The students can complete the activity on page 62 of their Workbook.

Graded activities

1 Ask the students to complete the diagrams in their Workbook on page 63. They should draw their own diagram to show the way that the Moon reflects light from the Sun.

2 Ask the students to use the ideas from the Students' Book to create their own comic strip story that helps to explain to younger children what light sources are. Display these in the classroom.

3 The students should do their own research on the questions about the Moon. They will need to use diagrams or a model of the Earth, the Sun and the Moon to do this. (They will learn more about this in Topic 4, but they should also recall some knowledge from earlier stages.)

Consolidate and review

- Let students work in pairs to draw diagrams to show how they can see an object. Let them choose a real object in the classroom, identify the light source and the way the light is reflected. Let them explain this to their partners. Walk around, listen and ask questions to make sure the students understand.
- Reinforce the students' learning in the lesson by showing Slideshow P5 about reflected light.

Differentiation

■ All of the students should be able to explain that light is reflected from surfaces and that this enables us to see things that are not light sources themselves.

● Most of the students should be able to create a comic strip about light sources without much assistance.

▲ Some of the students should be able to explain, using a diagram or a model, the way that the Moon reflects light from the Sun. Some students will also be able to explain why we sometimes see only part of the Moon or no Moon at all.

3.11 Reflecting and absorbing light

Student's Book pages 68–69

Physics learning objective

- Know that beams/rays of light can be reflected by surfaces including mirrors, and when reflected light enters our eyes we see the object.

Resources

- Workbook page 64
- DVD Activity P2: Reflecting light

Classroom equipment

- large dark-coloured basin or bowl half-full of water
- variety of materials with reflective or non-reflective surfaces, e.g. as listed on Workbook page 64
- small plastic mirrors
- distorting mirrors, if possible
- flashlights
- large sheets of paper to make posters
- coloured pens or pencils

Scientific enquiry skills

- *Plan investigative work:* Use knowledge and understanding to plan how to carry out a fair test; collect sufficient evidence to test an idea; identify factors that need to be taken into account in different contexts.
- *Obtain and present evidence:* Make relevant observations; discuss the need for repeated observations and measurements.
- *Consider evidence and approach:* Decide whether results support predictions.

Key words

- **absorb**
- **scatter**

⚠ Students should be careful with mirrors as they can break and cause injury. Remind them that light sources can get hot, and not to look directly at them.

Scientific background

Most surfaces reflect the light that falls on them by *scattering* it in all directions. However, shiny, lightly coloured, smooth and flat surfaces (such as mirrors and polished metal) reflect light in such a way that an image is formed. On these surfaces, the light bounces off at the same angle as it lands and this is why we see an exact image (although it is reversed horizontally). Some surfaces do not reflect light well. These surfaces are dark and uneven; they *absorb* much of the light.

Introduction

- Half-fill a dark-coloured basin with water. Invite students to come up and look at themselves in the water. Ask them what they can see and why they can see themselves. Then move the water about with your hand and ask them to look again. What do they see this time? Why? Tell students to think about rays of light and where they may be going.

Teaching and learning activities

- Show the students the selection of materials that you have brought to class. Ask them to predict whether each material will reflect light well or not. Ask: *What do you think will happen to the light when it lands on this surface? Will we be able to see an image clearly on the surface? Why?*
- Let students open their Student's Book and look at the photograph of the still lake. Discuss why we would be able to see reflections clearly in this water.
- Draw diagrams on the board to show how light is reflected off even and uneven surfaces. Show the students that the angles at which the rays hit the surface and bounce off again are the same on an even surface. Ask students to think where the light goes when it bounces off an uneven surface. Ask: *How does this affect what we see?*
- Let the students draw their own diagrams and discuss how light is reflected off different types of surfaces. Remind the students that light travels in straight lines.
- If you have any mirrors that distort images, show these to the students. Let them look at the surfaces and predict what images they will see. Then let them try out the mirrors and see if their predictions were correct.

Graded activities

1 Ask the students to give two examples of surfaces that reflect light well and two of surfaces that do not. They should also explain why the surfaces do or do not reflect the light so well. They can draw a simple diagram to show this and use it in their explanations.

2 Students work in groups and set up investigations to find out which materials reflect light well. Ask them what they will do to test this. Give them flashlights and small plastic mirrors, and a selection of shiny and dull surfaces. Let them shine the flashlight on the different surfaces. Ask: *Which surfaces reflect the flashlight? In which surfaces can you see yourself?* Encourage the students to position the surface being tested so that it is possible to see the beam of light on a sheet of paper as it travels to and from the surface. Then ask: *What would be a good way to write down our results to show to other people?* Encourage them to tabulate their results and draw diagrams to show the path of the light. Discuss the results with the class. The students should have found that flat, shiny surfaces are the most reflective and that these allow us to see ourselves most clearly. Discuss what happens in other cases: if the surface is uneven, the light can be reflected and travel away in a variety of directions, which breaks up the image. Some surfaces scatter the beam of light, and in these cases a coherent beam cannot be seen leaving the surface – no image is visible.

The students should follow the guidelines in their Workbooks on page 64, also recording the details of their investigation on that page.

3 Let the students choose two materials, one that reflects well and another that does not reflect so well. They should make a poster to show the differences and give scientific explanations for these. They should make diagrams and use data that they have collected from their investigation.

Consolidate and review

- Ask the students to explain the results of their investigation as a way of reviewing and consolidating what has been learned.
- Let the students complete DVD Activity P2 to consolidate what they have learned about reflected light.

Differentiation

■ All of the students should be able to give examples of surfaces that reflect light well and surfaces that do not reflect light well and say why this is so.

● Most of the students should be able to set up a simple investigation to test whether materials reflect light well or not. They should be able to gather suitable materials to test, record their predictions and then carry out their investigations and record the results. After that, they should be able to rank a range of materials according to how well they reflect light.

▲ Some of the students should be able to draw diagrams and give a clear and detailed explanation of why one material reflects light better than another material.

3.12 Changing the direction of light

Student's Book pages 70–71

Physics learning objective

- Explore why a beam of light changes direction when it is reflected from a surface.

Resources

- Workbook pages 65 and 66

Classroom equipment

- large mirror
- small plastic mirrors
- flashlights
- rulers and protractors
- black paper
- wooden blocks
- a periscope, if possible
- materials for building model periscopes, as listed on Workbook page 65 (optional; students could bring these in from home)

Scientific enquiry skills

- *Ideas and evidence:* Use observation and measurement to test predictions and make links.
- *Obtain and present evidence:* Make relevant observations; discuss the need for repeated observations and measurements.
- *Consider evidence and approach:* Decide whether results support predictions.

Key words

- **angle**
- **periscope**

If plastic mirrors are not available, mirror tiles can be cut to size as a safer alternative to glass mirrors. If glass mirrors are to be used, cover them with a strong clear adhesive sheet to reduce the dangers if they are smashed. Put tape round any sharp corners on the mirrors.

Scientific background

Most surfaces reflect the light that falls on them by scattering it in all directions. Shiny, flat surfaces (such as mirrors and polished metal) reflect light in such a way that an image is formed. The light ray reflects (bounces) off the surface at the same angle as it hit it. We can see 'round corners' when light has travelled from an object to our eyes via a reflective surface or mirror, for example in devices such as *periscopes*.

Students may be confused between shadows and images. In the formation of a shadow, an opaque object is blocking the light from a source. In the formation of an image, the light from an object is changing direction before entering the eye.

Students may have difficulty explaining why a mirror image is reversed left to right but not top to bottom. Encourage them to explore what happens when they move a part of themselves to touch the mirror: their right hand touches the mirror on their right (the mirror's left), but the top of their head touches the mirror at the top of the image's head.

Introduction

- Remind the students that they see objects because light reflected from the object enters their eyes. Tell the class to look towards the front. Ask: *Without moving your heads, can you see the people who are sitting next to the window (i.e. to the extreme right or left of the class)? Can you see the students who are sitting next to the wall? Why? Does the light reflected off those people reach your eyes?*
- Position the large mirror in front of the class and ask the students to look at it. Ask: *What can you see now? What is different? How does the light get to your eyes?* Help the students to verbalise what is happening to the light – it leaves a person, hits the mirror, reflects (or bounces off) and reaches another person's eyes, so that person can see the first person. Explain to the students that, more correctly, we say that we can see the person's image.

Teaching and learning activities

- Ask the students to open their Student's Books and look at the picture of the snooker player. Let them discuss what they see and observe the angles. Explain that light behaves in the same way as this ball on a snooker table. Light bounces off things and this is how it changes direction.
- Ask whether any students have seen film of a periscope in use in a submarine. Ask: *What is the periscope for? Can you suggest how it works?* If you have a periscope in class, demonstrate its use and allow students the opportunity to come and look through the periscope.
- Then let students work in pairs to discuss the picture of the girl with the periscope. They should trace the path of the light backwards, from the girl, over the wall and to the children on the other side of the wall. After students have worked in pairs, take feedback. If necessary, prompt the students to say that mirrors are used to make the light go in different directions, so that it can get to our eyes and we can see.
- Let the students work in pairs again. Give them two small mirrors and let them discuss a way to arrange them so that they can see the back of their own head. Once they have succeeded, let them draw diagrams to show how it was possible. They will need to show the directions of the rays of light and the way that the mirrors change the direction of the light rays, bouncing them in other directions.
- Students can build a periscope using the guidelines in their Workbook on page 65. They could, perhaps, build these at home and bring them to class. If there is time, let the students bring the materials they need from home and then build the periscopes in class. Once the periscopes are complete, the students should be able to demonstrate how they work, explaining briefly how the light rays travel.

Graded activities

For these activities, it is recommended that the students work in mixed ability groups. Differentiation should be via the level of support the students receive as they work on the activity, as well as by outcome (please see guidance in the 'Differentiation' box bottom right).

■ 1 Give the students a flashlight and a mirror and let them play with light and reflections. Walk around, asking them to explain where the rays of light are going.

● 2 Ask the students to investigate ways to get light to move around a corner. They should use their knowledge of how light moves and the way that it can be reflected to set up this investigation. Then they should draw a diagram in their Workbook on page 66. The diagrams should include accurate measurements.

▲ 3 Let the students play a game. They should use wooden blocks and black paper to make a race track. Then, using a flashlight and mirrors, they have to get the light to shine all along the track. Once they have succeeded, let them demonstrate how they have done this to the rest of the class.

Consolidate and review

- Use the periscope to review and consolidate what students have learned about light and the direction of travel of light rays.
- Students could also work in groups to draw up questions to ask each other. As always, check the questions and the answers to make sure the students are not passing on incorrect information.

Differentiation

■ All of the students should be able to explain that light travels in straight lines that do not bend, but that light can be reflected off surfaces so that we can see around corners. They should be able to demonstrate this with the aid of a flashlight and a mirror.

● Most of the students should be able to set up an investigation into the way that light can change direction and draw a diagram to explain what happens to the light.

▲ Some of the students should be able to set up a game in which they change the direction of light rays. They will understand and be able to explain to less able students what they have done to make this work.

Consolidation

Student's Book page 72

Physics learning objectives

- Observe that shadows are formed when light travelling from a source is blocked.
- Investigate how the size of a shadow is affected by the position of the object.
- Observe that shadows change in length and position throughout the day.
- Know that light intensity can be measured.
- Explore how opaque materials do not let light through and transparent materials let a lot of light through.
- Know that we see light sources because light from the source enters our eyes.
- Know that beams/rays of light can be reflected by surfaces including mirrors, and when reflected light enters our eyes we see the object.
- Explore why a beam of light changes direction when it is reflected from a surface.

Resources

- Slideshow P6: Light and shadows
- Assessment Sheets P1 and P2

Looking back

- Use the summary points on page 72 of the Student's Book to review the key things that the students have learned in this topic. Ask questions.
- Use Slideshow P6 about light and shadows to review the work on shadows.

How well do you remember?

You may use the revision and consolidation activities on Student's Book page 72 either as a test, or as a paired class activity. If you are using the activities as a test, have the students work on their own to complete the tasks in writing. Collect and mark the work. If you are using them as a class activity, you may prefer to let the students do the tasks orally, in pairs. Circulate as they discuss and answer the questions, observing the students carefully to see who is confident and who is unsure of the concepts.

Some suggested answers

1 Answer will vary. For example, the Sun, a lit candle, a flashlight.
2 Students should draw pictures that show the different angles and lengths of the shadows. The explanations should contain the following key ideas: shadows change because the Earth moves and so the Sun is not always in the same position in the sky. Shadows get shorter when the Sun is directly above us and longer when the Sun is rising or setting (in the morning and afternoon).
3 Students should draw a picture of the eye, straight light rays coming from an object, entering the eye at the pupil and an image forming on the retina.
4 Reflection; for seeing round corners.

Assessment

A more formal assessment of the students' understanding of the topic can be undertaken using Assessment Sheets P1 and P2. These can be completed in class or as a homework task.

Students following Cambridge International Examinations Primary Science Curriculum Framework will write progression tests set and supplied by Cambridge at this level, and feedback will be given regarding their achievement levels.

Assessment Sheet answers

Sheet P1

1 true / false / false / true / true [5]
2 light / blocked / position [3]
3 Transparent materials let a lot of light through. Translucent materials let some light through. [2]
4 false / true [2]
5 mirrors / reflect [2]
6 Diagram B. [1]

Sheet P2

1 Size, shape and direction. [3]
2 a light source / reflected off / reflected [3]
3 false / true / true / false [4]
4 Check students' diagrams. [1]
5 Check students' diagrams. [2]
6 light / reflected [2]

Student's Book answers

Pages 48–49

1 A shadow is a shape that is formed when an object blocks light. Blocking light causes shadows. Light travels in a straight line.
2 The light is blocked and a show is formed. It will be light where the light is hitting the tree trunk and it will be dark behind the trunk, in the shadow of the tree.
3 Answers will vary but might include: A shadow is a dark area where the light from a light source does not reach. When light is blocked, a shadow is formed. The length of the shadow will depend on the time of day. Shadows are longer in the morning and afternoon, and shorter in the middle of the day.
4 There would not be a shadow; there would be light.

Pages 50–51

1 The shadows in picture B are shorter than in picture A, and the shapes of the shadows are not the same. The position of the Sun in the sky is not the same – it is higher in the sky and more to the right in picture B.
2 They change in size, shape and direction.
3 Because of the apparent movement of the Sun across the sky. (Although it is actually the Earth that is moving, not the Sun.)
4 The shadows change in length, shape and direction, being shortest at noon.
5 Because of the changing direction of the Sun, and its height in the sky.

Pages 52–53

1 The shadow will get bigger.
2 The shadow will get smaller.
3 Position B.

Pages 54–55

1 The lines and numbers are not always evenly spaced, and some do not have a circle of numbers; the numbers on the vertical sundial are arranged anticlockwise.
2 A sundial has a stick in the middle that casts a shadow on the dial. We can work out the time from the position of the shadow on the sundial.
3 No, because there are no shadows at night.
4 The diagram shows the length and direction of the shadow at different times.

Pages 56–57

1 Opaque: you cannot see through it, it blocks all the light. Translucent: lets some light through. We can see through these materials a little, but we cannot see clearly. Transparent: allows light to pass through it. We can see through these materials.
2 Answers will vary. If the paper is very thin, it will probably be translucent; if it is heavy and thick it will be opaque.
3 The visor is transparent so that the person can see where he or she is going. The rest of the helmet is opaque and will block sunlight reaching the head (but that is not the main function of the material, which is to protect the head in the event of an impact).

Pages 58–59

1 A big lamp; it is behind the puppets.
2 The puppets are opaque so that they create shadows.
3 If the screen was opaque, the audience would not see the puppets. If the screen was transparent they would see the puppets being manipulated behind the screen.
4 Yes; it will be black as no light goes through opaque materials.
5 Yes; it will be fuzzy and coloured.

Pages 60–61

1 Answers will vary. Some classrooms are bright and others are dark/dull. The light outside is usually brighter during the day and in summer, but it may be dull if the weather is overcast.
2 About 400 inside the classroom, and 10 000–25 000 on the sports field.
3 Answers will vary. For example, to photographers who need to know the intensity of light in order to take clear photographs; to architects who need to measure light inside a building to see if it is adequate; to gardeners who grow plants artificially or indoors and who need to monitor the amount of light the plants are getting to ensure good growth.

Pages 62–63

1 Architects should know what the light intensity is inside a building and design its lighting so that the light level in it will be comfortable for people to work and live.

Pages 64–65

1 Students' own observations.
2 They stop (block) harmful light from entering the eyes.

Pages 66–67

1 There could be light coming from the Moon, a car or street lights. The light is reflected off surfaces towards our eyes.
2 Students should be able to say that it tells us that we can't see if there is no light and that many things reflect light.

Pages 68–69

1 Students may suggest that because the water is smooth and flat it reflects light well, like a mirror.
2 The diagrams show rays of light being reflected in a scattered, uneven way from an uneven surface, and in a regular way from a smooth surface.

Pages 70–71

1 There are mirrors inside the periscope.
2 The students need two mirrors to reflect the light in a way that allows them to see the back of their head – one mirror in front of them and one behind them. The light travels from the back of their head to the mirror at the back and then to the mirror in front and to their eyes.

4.1 Where does the Sun go at night?

Student's Book pages 74–75

Physics learning objective

- Explore, through modelling, that the Sun does not move; its *apparent* movement is caused by the Earth spinning on its axis.

Resources

- PCM P3: Making a model of the Earth
- PCM P4: Our spinning planet Earth
- Video P1: The movement of the Earth

Classroom equipment

- globe, or an orange with a stick through it, and flashlight
- orange ball
- materials for students to make papier mâché models of the Earth: balloons, glue, newspapers, paint and brushes, sticks for the axis (alternatively, use plastic or polystyrene balls)

Scientific enquiry skills

- *Ideas and evidence:* Know that scientists have combined evidence with creative thinking to suggest new ideas and explanations for phenomena.
- *Obtain and present evidence:* Make relevant observations.

Key words

- **sphere**
- **model**
- **horizon**

Remind students never to look directly at the Sun. Warn students to be careful with the sticks that they use to make the axis of the Earth and to clean up any spillages of glue or water which could make the floor slippery.

NOTE: If you choose to make papier mâché models of the Earth you will need to allow time for these to dry and set hard.

Scientific background

The Sun appears to move across the sky during the day. We say that it rises in the east and sets in the west. In fact, it is the Earth that moves. The Earth rotates (spins) on its axis (anti-clockwise when viewed looking down on the North Pole) and completes a full rotation every 24 hours. Day length varies with the different seasons. The variation is considerably greater in regions towards the poles than in regions close to the equator. Again, this is linked to the tilt of the Earth, which affects the regions towards the poles more than those at the equator.

Introduction

- Ask the students: Where does the Sun rise? Where does it set? Ask them to point and then ask for volunteers to come and draw the 'movement' of the Sun during a day. Help them to draw a curve that moves from east to west. Ask them if they know where east and west are and, if they don't, show them. Introduce the term *horizon*, telling the students that this is the line where the Earth's surface and the sky seem to meet.
- Ask: Does the Sun rise and set in the same place all year around? Some students may have noticed that the position changes.
- Show Video P1: The movement of the Earth

Teaching and learning activities

- Explain that the Sun does not move. It is actually the rotation of the Earth that makes it seem that way. Use a globe and a flashlight to demonstrate the Earth spinning anti-clockwise (west to east) rather than the Sun moving clockwise (east to west). (Alternatively, use an orange or a ball or other *sphere* with a stick stuck through it as a *model* of the Earth, so that the students focus on the movement first and not on the information on the globe.)
- Use the flashlight and globe model to track with the students the order in which regions of the world of different longitudes would witness sunrise (or sunset).
- Ask the students to look at the time-lapse photo on page 74 of the Student's Book. Ask them what they observe. Let the students work in pairs to discuss the answers to the questions. Take feedback and address any misconceptions.

- Further demonstrate the concept of the spinning Earth by asking the class to take part in another demonstration. Ask a student to hold up an orange ball to be the Sun. Use the globe to demonstrate that, in the daytime in their country, they are facing the Sun (the orange ball). Rotate the globe so that their country is facing away from the Sun. Explain that when their country is not facing the Sun, they get night-time and the other side of the world gets daytime. Repeat this a few times and ask the students to say 'daytime' when their country is facing the Sun and 'night-time' when it is facing away from the Sun. Reinforce that the opposite side of the world is having night-time when we have daytime.

Graded activities

■ 1 In groups, the students should build a papier mâché model of the Earth following the instructions on PCM P3. Explain that they need to look carefully at the globe because they are going to make their own models so that they can demonstrate the movement of the Earth. Ask them to discuss how they can make a model and what materials they will need to make it.

● 2 In their groups, the students should use their models to demonstrate that the Earth is spinning and the Sun is staying still, so that when one part of the Earth is facing the Sun it is daytime and on the other side it is night-time. They should use the ideas and methods from the earlier class demonstrations. Encourage the students to hold their model Earth at an angle, letting them look at the globe for reference.

▲ 3 The students should use their model to explore the idea: *Sunrise (and sunset) will not occur at exactly the same time in all parts of our country.* Let the students use a flashlight and shine it on their models as they turn them to see that sunrise will 'travel' across the country at a constant speed. You may need to explain that most countries use one standard time for the whole country because this is more convenient.

Consolidate and review

- Draw a picture of the Sun on the board. Let the students stand up and turn slowly on the spot towards and away from the fixed Sun, so they can see that when they face the Sun it would be daytime and when they face away from the Sun it would be night-time.
- In pairs, let the students use PCM P4 to reinforce that fact that it is the Earth that is spinning and not the Sun that is moving.

Differentiation

■ All of the students should be able to explain why Sun appears to move in the sky and state that it only *appears* to move, it is really the Earth that moves. They will be involved and engaged when making a model of the Earth. Most of the students will be aware that the stick (axis) needs to be aligned through the North and South Poles.

● Most of the students will be able to use their models to demonstrate that the Earth is spinning on its axis. Most will replicate the demonstrations already used in class. Some students may come up with new ideas of ways to model this.

▲ Some of the students will be able to give a clear and detailed explanation about why sunrise (and sunset) does not happen at the same time in different parts of a country.

4.2 The Earth rotates on its axis

Student's Book pages 76–77

Physics learning objectives

- Explore, through modelling, that the Sun does not move; its *apparent* movement is caused by the Earth spinning on its axis.
- Know that the Earth spins on its axis once in every 24 hours.

Resources

- Workbook pages 67–68
- Video P2: The rotating Earth

Classroom equipment

- students' Earth models from Unit 4.1
- globe (or an orange with a stick through it)
- flashlights

Scientific enquiry skills

- *Ideas and evidence:* Use observation and measurement to test predictions and make links.
- *Obtain and present evidence:* Make relevant observations.

Key words

- **axis**
- **rotate**
- **angle**

Scientific background

This unit introduces the word *axis* to the students. They will learn that it takes 24 hours for the Earth to *rotate* (spin round) once on its axis. This length of time is called a day. As the Earth spins on its axis, the part that is facing the Sun will be in the light (day). The part that is not facing the Sun will be in darkness (night).

The Earth's axis is tilted at an *angle* of 23.5°, so the Earth rotates at an angle. Some of the students will assume that the axis is vertical, so this tilt will need to be stressed.

Introduction

- Use the globe, a model or an orange with a stick through it and remind students that the Earth spins around. Teach the meaning of the word 'axis', if the students don't already know it. Make sure they understand that an axis is an imaginary line through the Earth. Ask them about other uses of this word; they may know the word from mathematics (an axis is an imaginary line that can divide things in half, and graphs have horizontal and vertical axes).
- Show the class Video P2 of the rotating Earth.
- Read through the information on pages 76–77 of the Student's Book. Refer students to the diagram on page 77 and make sure they understand what it is showing. Discuss the questions as a class.

Teaching and learning activities

- Teach the words 'rotate' and 'rotation'. Explain that the Earth rotates around its own axis and that it takes 24 hours to complete one full rotation – in other words, to turn around completely. Remind the students that it turns in an anti-clockwise direction.
- Then ask: *What has the Earth's rotation on its axis got to do with daylight and night-time?* The students should already have an idea about this from earlier demonstrations and investigations. Ask for volunteers to come up and explain. Ask: *Will the amount of daylight and night-time be the same or different in different parts of the Earth?* (It will be different.) Demonstrate this or ask students to demonstrate.
- Let students work in groups with their models. They should find the axis and turn the model around to make a complete rotation. Then let them use flashlights again to create daytime and night-time in different parts of the Earth. Let them compare the length of time the Earth takes to make one rotation with different meanings of the word 'day'. Ask: *Is there always the same amount of daytime?* (No.) (Students will learn more about the reasons for this in the next lesson. It has to do with the Earth orbiting the Sun.) *Is a day always 24 hours?* (Yes.) *Why?* (Because it takes the Earth 24 hours to turn on its own axis.)

Graded activities

1 All of the students should be able to read the statements on page 67 of their Workbook and say if they are true or false. Circulate and help any students who are struggling to read the statements. This activity will allow you to assess how well the students have understood the concept of the spinning Earth.

2 This is an individual exercise that most students should be able to undertake. Students should answer the questions on page 68 of their Workbook. Circulate and help any students who are struggling. This activity will allow you to assess how well the students have understood the concept of the spinning Earth.

3 Students should work in mixed ability groups using their models and flashlights. The students already know that the Earth moves around the Sun. Now they should think about the way the Earth is tilted on its axis and say what effect this angle has on life on Earth. Encourage them to think about the number of daylight hours and the temperature in different parts of the world. Ask them to think about ways in which things would be different if the Earth was not tilted. Ask: *Would we still have night and day?* (Yes.) *Would night and day be the same as now?* (No.)

Consolidate and review

- Let students work in small groups and make short presentations about what they have learned about the Earth and its axis. They should use a model in their presentation. Set a time for the presentation, for example two minutes. Each person in the group should participate, either saying something or demonstrating something.

Differentiation

■ All of the students should be able to answer the true or false questions with some help.

● Most of the students should be able to draw a simple diagram to illustrate the way the Earth turns on its own axis. They should be able to explain why this causes day and night on Earth.

▲ Some of the students will be able to explore and understand ways in which life on Earth would be different if the Earth were not tilted. More able students will be able to help other students elaborate and formulate these ideas, drawing from the facts and information that they already know.

4.3 The Earth's orbit

Student's Book pages 78–79

Physics learning objectives

- Explore, through modelling, that the Sun does not move; its *apparent* movement is caused by the Earth spinning on its axis.
- Know that the Earth spins on its axis once in every 24 hours.
- Know that the Earth takes a year to orbit the Sun, spinning as it goes.

Resources

- Workbook pages 69 and 70–71
- DVD Activity P3: The Earth's orbit

Classroom equipment

- students' Earth models from Unit 4.1
- globe (or an orange with a stick through it)
- flashlights

Scientific enquiry skills

- *Ideas and evidence:* Use observation and measurement to test predictions and make links.
- *Plan investigative work:* Make predictions of what will happen based on scientific knowledge and understanding, and suggest and communicate how to test these; collect sufficient data to test an idea.
- *Obtain and present evidence:* Make relevant observations; discuss the need for repeated observations and measurements; present results in bar charts and line graphs.
- *Consider evidence and approach:* Decide whether results support predictions; recognise and make predictions from patterns in data and suggest explanations using scientific knowledge and understanding.

Key words

- **orbit**
- **equator**

Remind students never to look directly at the Sun.

Scientific background

The Earth's axis is tilted (23.5° from the vertical, in relation to the plane of its orbit). As well as spinning, the Earth also orbits the Sun (once every 365.25 days). A good way to explain the effect of the tilting is to ask students whether daylight (and, therefore, night) would always be the same length if the axis was straight (it would be). They can then deduce that the tilt causes daylight length to vary.

As the Earth spins on its axis, causing daylight and night, the Earth is also orbiting the Sun in an oval-shaped (elliptical) orbit. At two points in the Earth's orbit – the spring and autumn equinoxes – the whole Earth experiences 12 hours of daylight and 12 hours of night-time. At these positions, no part of the Earth is tilted either towards or away from the Sun.

The Earth's orbit around the Sun takes 365.2425 solar days. (This is why the Gregorian calendar, which is a solar calendar, has a leap year every 4 years. The Islamic calendar or Hijri calendar is a lunar calendar, so dates are calculated differently.) One complete orbit of the Sun is a distance of around 940 million kilometres.

The distance from the Earth to the Sun varies during the orbit. The Earth moves at a speed of about 108 000 kilometres per hour.

Introduction

- Use the globe, a model or an orange with a stick through it to remind students that the Earth spins on its axis once every 24 hours while tilted at an angle of 23.5° from the vertical.
- Introduce and explain the word *orbit* to the students. Say that the Earth orbits the Sun.
- Ask the students to look at the diagram on page 78 of the Student's Book. Explain that, as the Earth rotates (spins), it also goes around the Sun in an oval-shaped orbit.

Teaching and learning activities

- Give the students a brief explanation and demonstration of the way that the Earth moves in an orbit around the Sun. Tell the students that the orbit is oval-shaped.

- Talk about how long it takes for the Earth to orbit the Sun. Ask: *How many days are there in a year? How long does it take the Earth to orbit the Sun?* Explain that the Earth takes a little more than 365 days to orbit the Earth and so every four years we have a leap year, so that the calendar matches the actual time it takes (and we can calculate years correctly).
- Ask the students to think about ways in which the movement of the Earth affects life on Earth. They will know about daytime and night-time and, from their work in Unit 4.2.

Graded activities

If appropriate, it is recommended that activity 1 is conducted as a whole class activity and is undertaken over a longer time period.

■1 This activity will need to be completed over a period of several weeks. Students should predict whether the Sun will be in the same place in the sky each time they observe it at intervals of one week. Choose somewhere where is a clear view of the eastern sky. Encourage the students to say how they can make this a fair test. (Observe the position of the Sun at the same time of day each time. Always stand in exactly the same spot.) They should record their findings on Workbook page 69. The observations prove that the Earth is moving around the Sun (orbiting the Sun). For more dramatic results you could make observations every two or four weeks, depending on the length of your school terms.

●2 The students can work in small groups. They should use their models of the Earth and a flashlight to explore how the tilt of the Earth affects the hours of daylight in a particular country. Let the students take turns to demonstrate their modelling to the class to show what they have learned. Encourage the groups to take part in a short question and answer session, you could use prompt questions to help them, such as: *If it were daytime in [name of country] would it be daytime or night-time in [name of country]? Which country do you think would have the most number of daylight hours?* Let the students use their models to help them to find the answer.

▲3 Students can work in small groups and look at the data on page 70 of their Workbooks. Explain that the tables give information about times of sunrise and sunset in Cape Town, South Africa, and in Tunisia in North Africa on the same days. Let the students find South Africa and Tunisia on the globe and describe where they are in relation to each other. Let the students study and discuss the data, thinking about the answer to question 2. Take feedback and explain if necessary. Then let the group discuss what type of graph we could draw to show the data. Each student should then work individually to draw a graph on page 71 of their Workbook. Afterwards, let them compare their graphs and decide on the most appropriate representation.

Consolidate and review

- Let the students use the globe and a flashlight to demonstrate the way that the Sun shines on different parts of the Earth at different times.
- Let the students complete DVD Activity P3 to consolidate what they have learned about the Earth's orbit.

Differentiation

■ All of the students should be able to complete this activity with little or no help. Less able students may need extra support when making their predictions. Students should be able to record basic data to show the position of the Sun. Most of the students should also be able to interpret data with the help of peers in the group.

● Most of the students should be able to model how the tilt of the Earth affects the hours of daylight. Be prepared to help any groups that find this concept difficult. The students should be able to take part in a class question and answer session, more able students will be able to confidently lead the discussion.

▲ Some of the students will be able to interpret data, make a graph of the data and draw conclusions with little assistance. Most of the students will be able to complete this activity with some help. Less able students will find this a challenging activity and it may be better to work through the data with them, helping them to construct a graph.

4.4 The Solar System

Student's Book pages 80–81

Physics learning objective

- Research the lives and discoveries of scientists who explored the Solar System and stars.

Resources

- Workbook page 72
- PCMs P5–P12: Planet outline templates
- PCMs P13–16: Planet fact cards
- Slideshow P7: Planets in the Solar System
- DVD Activity P4: The planets
- Video P3: The planets

Classroom equipment

- coloured pens or pencils
- long measuring tape and sticky pads
- access to the internet or reference books
- large sheets of paper for making posters
- printed information and facts about Ursa Major and Orion, if necessary

Scientific enquiry skills

- *Ideas and evidence:* Know that scientists have combined evidence with creative thinking to suggest new ideas and explanations for phenomena.
- *Obtain and present evidence*: Present results in bar charts and line graphs.

Key words

- **planet**
- **moon**
- **Solar System**
- **constellation**

NOTE: Before the lesson, research some suitable and safe websites for the students to refer to when researching planets and constellations. The NASA website has lots of good and accurate information.

Scientific background

The *Solar System* consists of eight *planets*: Mercury, Venus, Earth, Mars, Jupiter, Saturn, Uranus, Neptune. Some of the planets are themselves orbited by one or more *moons*. There is a clear distinction in size between the first four and the last four planets as you move away from the Sun. Each of the planets orbits the Sun, with the orbit time increasing as distance from the Sun increases. The time taken for a planet to orbit the Sun is known as its year. Illustrations of the Solar System in books can be misleading, as the distances of the planets from the Sun cannot be shown to scale.

The Sun is just one of billions of stars in the Universe. All stars are light sources; they give out light, whereas the planets and the Moon are reflectors of light. Stars are always present in the sky but we only see them at night, when the sky is not filled with light from the Sun. Stars also become visible during a total solar eclipse.

Constellations are patterns made by groups of stars that appear in the night sky. In reality, the stars in a particular constellation are not close together in space; they only appear to be grouped. Some are much more distant than others.

Introduction

- Use the diagram of the Solar System on page 80 of the Student's Book to remind the students that the Earth moves around the Sun. Introduce the words 'Solar System', 'planet' and 'orbit' as you do this.
- Discuss question 1 as a class. Some students may already have a handy mnemonic to remind them of the names and order of the planets – ask them to share these with the class. A well known mnemonic you could teach the students is: My Very Educated Mother Just Served Us Nachos.
- Stress that all the planets move in orbits around the Sun. You can also mention moons, which move in orbits around the planets.

The students do not need to learn about the Solar System or the planets as part of Cambridge Primary Science curriculum for Stage 5. This lesson goes a little beyond the requirements of the curriculum and is included here for enrichment purposes and to place the objective within a more familiar context.

Teaching and learning activities

- Show Slideshow P7 about the planets in the Solar System. Use your own narration if you judge the English too difficult for your students to follow easily.

- Ask the students to use their knowledge of the Solar System: *Which planet is the hottest?* (Mercury.) *Which planet is the coldest?* (Neptune.) Ask: *Why do they think this?* (Distance from the Sun.)
- Let the students complete the activity on page 72 of their Workbook to create a bar chart comparing how fast the different planets move around the Sun. Explain that one orbit around the Sun equals one year on that planet.

Graded activities

1 Give each group of students one of the planet outlines from PCMs P5–P12. They should colour in their planet using the diagram in the Student's Book, or any other sources, as a guide. Then get the students to create a display on one wall of the classroom. Have the 'Sun' at one end and ask the students to stick their planet at the correct relative distance away from the Sun, to give a representation of the distances between the planets. Let the students do some research to find out about the moons that orbit their planet and then add these to the display (just a representative sample will be fine, as the exact number of moons orbiting some of the planets is still being debated by scientists).

2 Let the students choose a constellation to research. They can use the internet or books and draw a star map of it. Let them re-tell or act out to the rest of the class any stories associated with their chosen constellation. If you do not have classroom access to the internet, print out some facts about Ursa Major and Orion prior to the lesson for the students to use.

3 Give each group of students one of the planet fact cards from PCMs 13–16. Let them use these and the internet to research more information about their planet. They should find out who first discovered it and when. Each group should make a poster to present to the rest of the class, showing what they have discovered. The posters can be added to the class display.

Consolidate and review

- Let the students complete DVD Activity P4, in which they place the planets in the correct order around the Sun.
- Show the students Video P3, which is an animation of the planets orbiting the Sun.
- Ask the students to look at the night sky and to try to name any of the constellations they can see. Depending on the time of year and where you are in the world, it is sometimes also possible to see some of the planets without a telescope.

Differentiation

■ All of the students should be able to correctly state the sequence of the planets. Most of the students will need some help to establish where their planet should be placed in relation to the Sun. Some of the students will be able to place their planet at the correct relative distance from the Sun. Less able students may need help working with 'millions' of years.

● Most of the students should be able to recognise and name a familiar star constellation. They will be able to undertake independent research to draw an accurate star map. Some of the students will realise that in reality, the stars are not grouped as they appear. Some are much more distant than others.

▲ Some of the students should be able to independently research facts about their planet to create an informative and well presented poster. Most students will be engaged and contribute to the group work. Less able students may need extra help with their research.

Big Cat

Students who have read *Big Cat Let's go to Mars* will recall the information about this planet in the Solar System.

4.5 Early astronomers and discoveries

Student's Book pages 82–83

Physics learning objective

- Research the lives and discoveries of scientists who explored the Solar System and stars.

Resources

- Workbook page 73 and 74–75
- Slideshow P8: The history of astronomy

Classroom equipment

- telescope to show to the students, if possible
- access to the internet and reference books
- presentation software (optional)
- rubber bungs and string (optional)

Scientific enquiry skills

- *Ideas and evidence:* Know that scientists have combined evidence with creative thinking to suggest new ideas and explanations for phenomena.

Key words

- **astronomy**
- **theory**
- **gravity**
- **Universe**
- **telescope**

NOTES: Before the lesson you should research the lives and discoveries of some of the early astronomers so that you are able to answer any questions the students might have. Also, research some suitable and safe websites for the students to refer to when researching information for the activity tasks. The NASA website has lots of good and accurate information.

Scientific background

In the past, it was widely believed that the Earth was the centre of the *Universe*. However, scientists such as Ibn al-Haytham (Alhazen) pointed out contradictions between the mathematics and the observations. During the 13th century, an observatory was built in Iran for the famous astronomer Nasir al-Din al-Tusi. He used his observations to calculate the length of a year accurately to within six decimal places. Since then, bigger and better *telescopes* have been produced, we send space probes to take pictures of space and robots to explore planets. We have space telescopes in orbit high above the Earth's atmosphere (such as the Hubble Space Telescope) and can send scientists into space.

Introduction

- Ask the students how we know about the Solar System. They may describe the use of telescopes, visits to the Moon or images from space probes.
- Show Slideshow P8 to explain the history of *astronomy*.
- Explain that there is a lot of information about the Solar System and the history of astronomy. You can focus on areas that interest your students most, after you have covered the basic information. The main focus in this unit should be on the students doing their own research.

Teaching and learning activities

- Tell the students that, even in ancient times, people had *theories* about the Universe and the objects they saw in the sky. Explain that ancient people realised the importance of the Sun, Moon and stars. They also understood that there are patterns in the way the stars move across the sky; they started to use these to find their way, to understand directions, and to predict seasons and weather on Earth. They used them to decide when to plant or harvest their crops.
- Let the students read the information in their Student's Book and complete the questions.
- Ask if the students can think of any inventions that have helped astronomers to discover new planets. Then talk about how some theories, observations and ideas were proved when telescopes were invented. If possible, show the class a telescope and explain why they were so important in the development of scientific knowledge about the Solar System.
- Tell the students about the main astronomers in history: early Chinese astronomers, Nasir al-Din al-Tusi, Copernicus, Galileo, Newton, Halley, etc. But tell them that the list of scientists who have contributed to astronomy is huge. Explain that Islamic astronomers contributed many observations, ideas and theories, particularly between the 11th and 15th centuries. If culturally appropriate, mention to the students that, in the

past, scientists' theories about space have caused disputes with prevailing religions and that scientists were not always free to say what they found or believed to be true.

Graded activities

1 The students should undertake their own research, using the internet and reference books to find out how many moons Jupiter has and to explain why Galileo could not see them all with his telescope. (Current thinking is that there are at least 63. The largest are Io, Europa, Ganymede and Callisto.)

2 The students should use the information in their Student's Book, information from the lesson and their own research to compile a timeline of important discoveries in astronomy on page 73 of their Workbook.

3 This activity is for more able students. They should choose one of the scientists mentioned in the Student's Book (or lesson) and do their own research to find out about their life and discoveries. If possible, they should present their research to the rest of the class as a slideshow or digital presentation.

Consolidate and review

- *Gravity* is not studied until Stage 6 Topic 4 of this course, but it is a good idea to briefly review what the students have learned about gravity through their work to date on space and the planets. In very simple terms, gravity is the force that makes things fall when they are dropped. On Earth, it is the force that pulls (attracts) objects towards the centre of the Earth. The more massive an object is, the stronger its gravitational pull. Gravity is what holds the planets in orbit around the Sun and what keeps the Moon in orbit around the Earth. If you want to explore this further, you can use the activity on pages 74–75 of the Workbook to model the gravitational pull that holds two different planets in orbit.

Differentiation

■ All of the students should be able to find out some information about Jupiter's moons. Less able students may need additional help in order to read the information on English language websites or in books. The students should explain that Jupiter has many moons, and that some of them are very small so that Galileo's telescope would not have been powerful enough to see them.

● Most of the students will be able to use the information in the Student's Book and Workbook to create a timeline. Through independent research, the more able students will include the names of astronomers (and their discoveries) that you have not specifically covered during the lessons.

▲ More able students should find this research more of a challenge. They will need to do independent research and produce a presentation to show the rest of the class. Their presentation should be informative and clear.

4.6 Space exploration today

Student's Book pages 84–85

Physics learning objective

- Research the lives and discoveries of scientists who explored the Solar System and stars.

Resources

- Workbook page 76

Classroom equipment

- access to the internet and reference books

Scientific enquiry skills

- *Ideas and evidence:* Know that scientists have combined evidence with creative thinking to suggest new ideas and explanations for phenomena.

Key words

- **satellite**
- **space station**
- **astronaut**

NOTE: Before the lesson you should research the lives and discoveries of some modern astronomers so that you are able to answer any questions the students might have. Also, research some suitable and safe websites for the students to refer to when researching information for the activity tasks. The NASA website has lots of good and accurate information.

Scientific background

On 20 July 1969, Neil Armstrong became the first man to walk on the Moon. He and Buzz Aldrin explored the Moon for less than three hours. They returned to Earth with 22 kg of rocks for scientists to study. NASA had carried out ten Apollo Moon missions before Neil Armstrong landed on the Moon in Apollo 11. Some of these missions were unmanned, and some of them landed on the Moon. There were also some manned missions that orbited very close to the Moon's surface, but the *astronauts* did not get out.

The International *Space Station* (ISS) is a habitable artificial *satellite* in orbit around the Earth. It is the ninth space station to be inhabited. The ISS is in orbit approximately 350 km above the surface of the Earth. It is a collaborative project, built and used by many different countries. The ISS was put together from many different pieces that were sent into space and then assembled there. The first piece was launched in 1998 and the station was completed in 2011. Astronauts have been living there since 2011. The ISS can house six astronauts and is about the same size as a five-bedroomed house. The role of the ISS is to carry out experiments and research. It is a science laboratory, as well as a home for the astronauts. Scientists are using the ISS to find out how to work and live in space.

Introduction

- Ask the students if they can name any present-day astronomers or space scientists. They may mention Professor Stephen Hawking or some of the NASA astronauts.
- Then tell the students about some of the modern developments in space exploration and stress that we are still learning about the Solar System. Talk about the American, Russian, Chinese and Indian space programmes, the Hubble telescope and the SKA programme. Select a few of the modern scientists and astronauts (such as Professor Stephen Hawking or the ISS astronaut Chris Hadfield) who have made a contribution to our knowledge, and encourage the students to do their own reading and research about them.
- Discuss any developments to do with space that feature in current news reports in your country.

Let the students read the information in their Student's Book pages 84–85 and complete questions 1 to 3 as a class.

Knowledge of satellites, (natural and human-made), is not required as part of Cambridge Primary Science curriculum and it will not be tested in the Progression or Primary Checkpoint tests. It is included here for enrichment and information purposes only.

Teaching and learning activities

- Explain that a satellite can be natural or artificial. The ISS and communications satellites are examples of artificial satellites. A natural satellite is a celestial body that orbits a planet, such as a moon. The Moon is an example of a natural satellite.
- Ask: *Why do we need to put artificial satellites into space?* Explain that artificial satellites affect our lives in many ways every day. They can broadcast television signals directly into our homes, they can help us to navigate through GPS or SatNav (Navstar Global Positioning Systems), they can provide meteorologists with information about weather systems, and they allow businesses to communicate with each other across the globe. Ask: *What would life be like if we did not have satellites?*
- Briefly talk about natural satellites. These include moons, which orbit the planets of the Solar System and beyond.
- Debate question 4 in the Students' Book with the class. Ask: *Do you think it is important to find out more about the Solar System?* There are no right or wrong answers here, but the question allows the students to think freely about how useful space exploration might be to their lives now and in the future. The students will be learning more about the future of space exploration in the next unit.

Graded activities

1 The students should find out about the work being conducted on the ISS and, in particular, why scientists want to know how the human body is affected by living in space for long periods of time. They should discover that it is because, if we want to send humans to explore other planets in the Solar System, it will take them many years to reach them. This means they will be living in space for very long periods of time. Encourage the students to find out about how muscle wastage occurs in zero gravity and why this might be a problem for humans.

2 Let the students undertake their own research to answer the questions on page 76 of their Workbooks.

3 The students should do their own research on modern day space programmes. This can be done individually or in groups. They could research the work done by the Indian, Chinese, American or Russian space programmes and produce a report detailing their major discoveries.

Consolidate and review

- Revise some of the most important discoveries made about the Solar System and stars, and the names of the scientists who made them.

Differentiation

■ All of the students should be able to find out some information about the work of the ISS. Less able students may need additional help in order to read the information on English language websites or in books.

● Most of the students will be able to undertake independent research. Less able students will need extra help.

▲ More able students should find this research more of a challenge. They will be able to do independent research and produce a report that is informative and well structured.

4.7 Into the future

Student's Book pages 86–87

Physics learning objective

- Research the lives and discoveries of scientists who explored the Solar System and stars.

Resources

- PCM P17: Orbits

Classroom equipment

- access to the internet and reference books
- pens and paper for timeline
- wooden or cork board, two pins, loops of string of two different lengths, large sheets of paper

Scientific enquiry skills

- *Ideas and evidence:* Know that scientists have combined evidence with creative thinking to suggest new ideas and explanations for phenomena.

Key words

- **asteroid**
- **comet**

NOTE: Before the lesson, research some suitable and safe websites for the students to refer to when researching planets and constellations. The NASA website has lots of good and accurate information.

Scientific background

Mars has already been visited by a number of NASA Mars Rovers in successful (and unsuccessful) robotic missions. These robots gather samples and record important scientific data for scientists back on Earth to study. Mars has a very thin atmosphere, consisting mainly of carbon dioxide. It is very cold (ranging from –120 °C to 25 °C). The distance from Earth to Mars varies greatly, due to the different orbits of Earth and Mars. As Earth is closer to the Sun than Mars, it completes its orbit more quickly. When Earth and Mars are at their closest, they are approximately 54 million kilometres apart. When they are at their furthest, they are approximately 400 million kilometres apart. If a mission to Mars was undertaken while Earth and Mars were at their closest, it is estimated that it would take at least 150 days to get there.

Comets are icy bodies that orbit the Sun, they have orbits that reach far beyond Pluto. It is thought that they may be waste material from when the outer planets formed. They contain dust, as well as ice. A comet is distinguished in the sky by its bright 'tail'. This is composed of gases that burn as the comet gets close enough to the Sun. Comets range in diameter from approximately 300 m to 10 000 km.

Asteroids are sub-planet sized rocks that orbit the Sun. Their small mass means they do not have enough gravity to pull themselves into a spherical shape. Some are almost as large as dwarf planets or moons, while others may be as small as a grain of sand. There is an asteroid belt between the inner and outer planets.

Over twenty large asteroids have been discovered so far, but there are millions of smaller ones. Asteroids are found in an inner belt and an outer belt. The inner belt contains asteroids made of metals. The outer belt contains asteroids made of different rock.

Introduction

- Ask the students what they can remember about the planet Mars. Refer them to the class display that they made in Unit 4.4 so that they can re-read the planet facts there, and to look at how far Mars is from the Earth and the Sun.
- Talk about how, one day, scientists would like to send humans to explore Mars. Ask: *Do you think that humans can breathe the air on Mars?* (No.) Explain that having a manned space station on Mars may one day be possible, but we would still need to have food, oxygen and water. It is long way to send supplies, and this would be very expensive – so it would be better if we could grow our own food on Mars. Ask: *What is one of the missions of the Mars Curiosity Rover?* (To find water on the surface of Mars.) *Why might this be useful?* (If there were ever a manned space station on Mars, there would be a local source of water for humans to use.)
- Let the students read the information on page 86 of the Student's Book and answer questions 1 and 2 as a class.

Teaching and learning activities

- Lead the class on to a discussion about the Earth. Ask: *What would happen if we used up all of the natural resources on the Earth?* (We would be unable to survive.) Suggest that other planets may contain natural resources that we could use instead. In the future, it might be possible to exploit the natural resources of other planets. Ask: *How does this make you feel?*
- Let the students read the information about comets and asteroids on page 87 of the Student's Book and answer question 3 as a class.
- Talk about how, in the future, scientists and industrialists believe that it will be possible to mine asteroids. Ask: *How practical do you think this would be?* Allow the students time to discuss this.

Graded activities

■1 Ask the students to imagine that they are in charge of controlling the Mars Rover. They should describe how they would use it to find out more about the planet. Let them discuss this within their groups.

●2 Ask the students to research the history of the exploration of Mars, from the invention of the telescope through to the modern day. They should use the internet and books to do this. In their groups, let them create a timeline of events to present to the class. They should compare the events that different groups have included on their timelines to see if they have the same events listed.

▲3 In groups, students should compile a list of reasons for and against the mining of asteroids and comets. They should consider things such as how much it would cost, how practical it is and how ethical it is.

Consolidate and review

- Let the students complete the activity on PCM P17 to compare two different-shaped orbits.

Differentiation

■ All of the students should be engaged and involved in the group discussion. They will all have their own opinions and ideas as to what they would want the Mars Rover to explore. All of the students should be able to listen to one another's ideas and take turns to speak.

● Most of the students will be able to do their own research with little or no help. All of the students should be able to contribute ideas for the timeline and be able to confidently peer review the work in their groups.

▲ Some of the students will find this activity challenges them to think theoretically about advances in technology that might happen in the future. More able students should be able to guide others when compiling the lists of pros and cons, giving sound reasons for why they have selected them.

Consolidation

Student's Book page 88

Physics learning objectives

- Explore, through modelling, that the Sun does not move; its *apparent* movement is caused by the Earth spinning on its axis.
- Know that the Earth spins on its axis once in every 24 hours.
- Know that the Earth takes a year to orbit the Sun, spinning as it goes.
- Research the lives and discoveries of scientists who explored the Solar System and stars.

Resources

- Assessment Sheets P3 and P4

Looking back

- Use the summary points on page 88 of the Student's Book to review the key things that the students have learned in this topic
- Let different groups use their model Earth from Unit 4.1 to demonstrate the way that the Earth moves around the Sun.

Distribute some coloured pens and ask the students to draw the planet of their choice. When they have completed their drawings, the students should write two facts about the planet.

How well do you remember?

You may use the revision and consolidation activities on Student's Book page 88 either as a test, or as a paired class activity. If you are using the activities as a test, have the students work on their own to complete the tasks in writing. Collect and mark the work. If you are using them as a class activity, you may prefer to let the students do the tasks orally, in pairs. Circulate as they discuss to answer questions, observing the students carefully to see who is confident and who is unsure of the concepts.

Some suggested answers

1. The Sun appears to move across the sky, but the Earth is actually moving around the Sun and rotating.
2. Students' own answers.
3. Students should be able to draw a diagram that shows the Earth orbiting the Sun in an oval-shaped orbit, tilted on its axis and rotating about its axis.
4. Telescopes enabled astronomers to make direct observations of planets and other bodies in space; they allowed astronomers to prove, by observation, some theories that had been devised through calculation, and to disprove others.
5. Students' own answers.

Assessment

A more formal assessment of the students' understanding of the topic can be undertaken using Assessment Sheets P3 and P4. These can be completed in class or as a homework task.

Students following Cambridge International Examinations Primary Science Curriculum Framework will write progression tests set and supplied by Cambridge at this level, and feedback will be given regarding their achievement levels.

Assessment Sheet answers

Sheet P3

1. false / true / false [3]
2. axis / hours / orbit / days [4]
3. sphere / Sun / anti-clockwise / spins [4]
4. Answers will vary but should include the following points: we have daytime and night-time because the Earth spins around [1]; When one part of the Earth faces the Sun, it is daytime on that part of the Earth and night-time on the part of Earth that is not facing the Sun. [2]
5. Answers will vary: accept any information that the students have researched and studied. [2]

Sheet P4

1. Check students' diagrams. [9]
2. false / true [1]
3. true / false / true [3]
4. Answers will vary but may include: collecting weather data and for communications. [2]

Student's Book answers

Pages 74–75

1 Because the Earth is spinning on its axis.
2 A sphere is a 3D ball shape.
3 clockwise
4 The Sun is still in the same place at night: it is the Earth that has rotated so that people on the opposite side are now the ones who can see the Sun.
5 You would be facing the Sun.

Pages 76–77

1 7 turns
2 The side facing the Sun would have daytime all of the time and the side facing away would be in darkness all of the time.
3 23.5°
4 The days and nights would be shorter (12 hour rotation) or longer (36 hour rotation).

Pages 78–79

1 It spins around its axis and it moves in an orbit around the Sun.
2 The Sun rises and sets at different times at different places on the Earth; it causes day and night.

Pages 80–81

1 Students' own answers.
2 It is warm enough for life; there is liquid water at its surface; it has an atmosphere.
3 The farmers could predict the different seasons and know when to expect the temperature to get warmer or cooler.

Pages 82–83

1 A scientist who studies the stars and planets.
2 Answers will vary but will probably include five scientists studied in the lesson.
3 The Earth orbits the Sun and not vice versa.

Pages 84–85

1 An object that orbits a planet.
2 Artificial satellites have been put into space by humans. Natural satellites occur naturally, for example moons.
3 Yuri Gagarin, in 1961.
4 Students' own answers.

Pages 86–87

1 Answers will vary, but could include: it is too far for humans to travel; humans could not survive when they reached Mars; a rover does not have to return to Earth; a rover is less expensive to send.
2 To discover whether it has any natural resources that humans could exploit.
3 The Earth will eventually run out of natural resources because humans are using them too up quickly. Mining asteroids and comets would allow humans to get more natural resources.

PCM B1: Do leaves need sunlight?

You will need:

- a plant with several big green leaves
- some thick, dark-coloured paper or cardboard
- scissors
- tape

1 Cut out a piece of paper that is big enough to cover one of the leaves of the plant.

2 Cover the leaf with the paper. Use the tape to keep the paper secure.

3 Leave the plant in a sunny place for at least a week. Water the plant as usual.

4 After one week, take the paper off the leaf and compare the leaf that had been covered with the other leaves on the plant.

5 Discuss why the leaves are different.

PCM B2: Making new plants

PCM B3: Bees

Read the passage below and then answer the questions.

Bees land on flowers to suck nectar. They are attracted by the scent and colour of the flower. The bees store some of the nectar inside them in a 'honey sac'. They fly back to the hive and use the nectar to make honey. The honey feeds the colony of bees. Some bees also use collected pollen to make honey.

Bees go from flower to flower, and pollen grains stick to their hairy bodies. Some pollen grains rub off on the next flower as the bee travels to each one. This is called pollination.

Bees and other insects pollinate plants that make the flowers that make the fruit we eat. If there are not many bees, less fruit is made. If there are lots of bees, more fruit is made. The number of honey bees is getting smaller. If bee numbers keep getting smaller, less fruit will be made. Fruit growers will suffer as they will not have fruit to sell.

1 What do flowers do for the bees?

____

____

2 What do bees do for the flowers?

____

____

3 Bees pollinate fruit trees. Why is this important?

____

____

4 If bee numbers drop, in what ways will fruit growers be affected?

____

____

PCM B4: What do seeds need?

1 Write down what seeds need to germinate.

____

2 Think about the jars that you set up in your investigation. Write down the variables that you were testing in this experiment. What was different in each jar?

____

____

3 Which of the jars A to D were missing something that a seed needs to germinate? Fill out the table below.

	What is missing for seed germination?
A	
B	
C	
D	

4 Predict which seeds will germinate the best. Why?

____

____

5 Which seeds do you think will grow the best once they have germinated? Why?

____

____

6 Write down what you did to keep the investigation fair.

____

____

PCM B5: The life cycle of a tomato

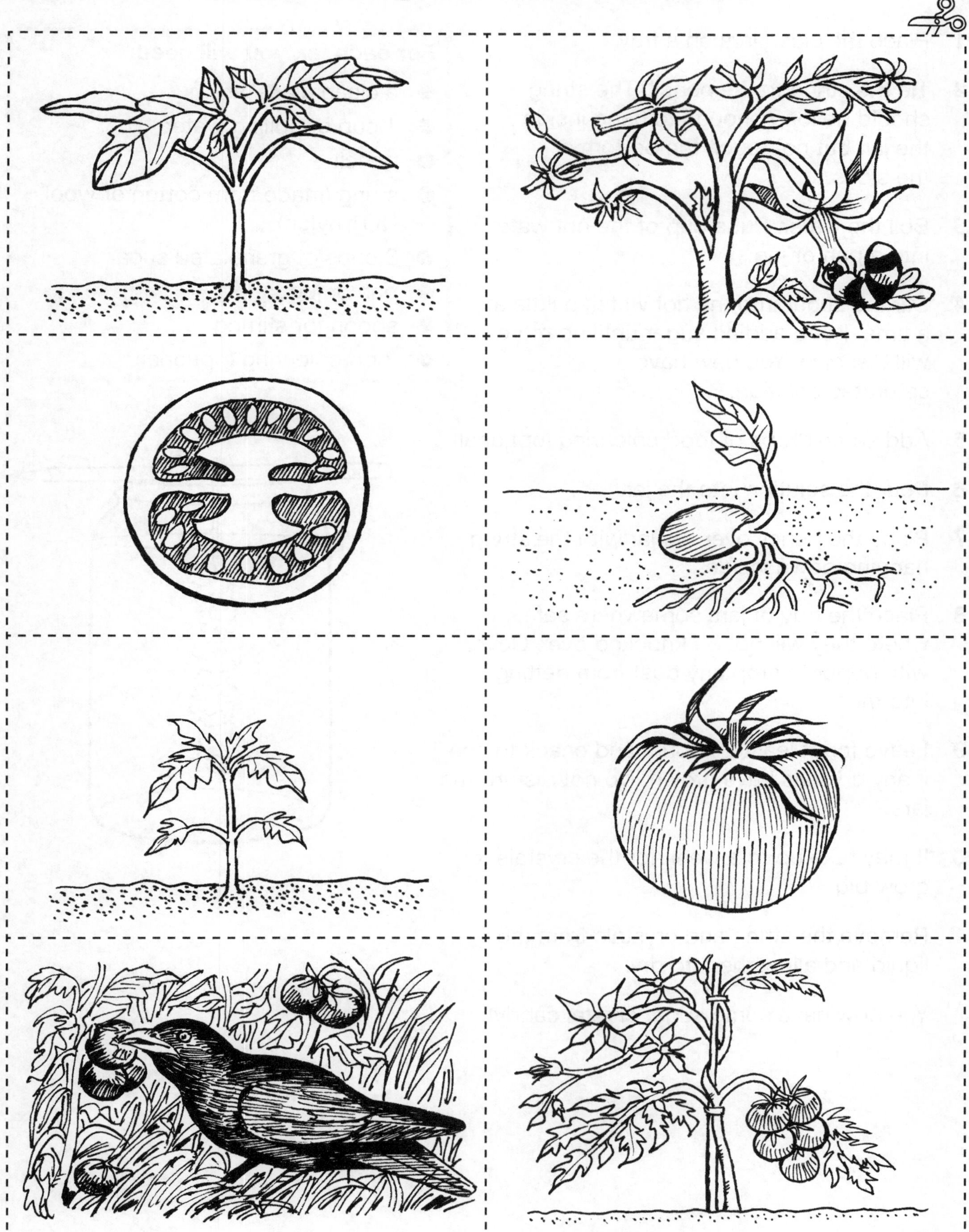

PCM C1: Making sugar crystal candy

For each jar, you will need:

- a clean glass jar
- 1 cup of boiling water
- pencil
- string (made from cotton or wool, NOT nylon)
- 3 cups of granulated sugar
- pan or jug
- spoon for stirring
- food colouring (optional)

1. Place the clean jars on a tray.
2. Tie the string to the pencil. The string should be long enough to hang inside the jar, but not touch the bottom or the sides.
3. Boil the water. Put a cup of the hot water into a pan or jug.
4. Stir the sugar into the hot water, a little at a time. Keep adding sugar until no more will dissolve. You now have a saturated solution.
5. Add some drops of food colouring (optional).
6. Pour the solution into the jar.
7. Place the pencil over the jar with the string hanging down inside.
8. Place the tray of jars somewhere safe where they will not be knocked over. Cover with paper to stop any dust from getting into them.
9. Leave for at least 24 hours and check to see if any crystals have grown. Do not disturb the jars.
10. It may take up to a week for the crystals to grow big.
11. Remove the string and crystals from the liquid and allow them to dry.

    You now have some sugar crystal candy!

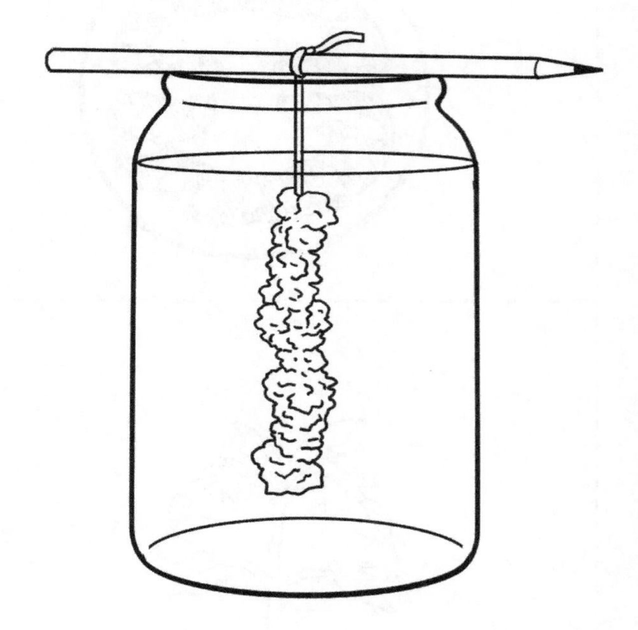

PCM C2: Making salt crystals

For each jar, you will need:

- a clean jar
- table salt
- boiling water
- pan or jug
- spoon for stirring
- pencil
- string (made from cotton or wool, NOT nylon)

1. Place the clean jars on a tray.
2. Tie the string to the pencil. The string should be long enough to hang inside the jar, but not touch the bottom or the sides.
3. Boil the water. Put a cup of the hot water into a pan or jug.
4. Stir the salt into the hot water, a little at a time. Keep adding salt until no more will dissolve. You now have a saturated solution.
5. Pour the solution into the jar.
6. Place the pencil over the jar with the string hanging down inside.
7. Place the tray of jars somewhere safe where they will not be knocked over. Cover with paper to stop any dust from getting into them.
8. Leave for a few hours and check to see if any crystals have grown. Do not disturb the jars.
9. It may take a few days for the crystals to grow big.
10. Remove the string and crystals from the liquid and allow them to dry.

You have grown some salt crystals!

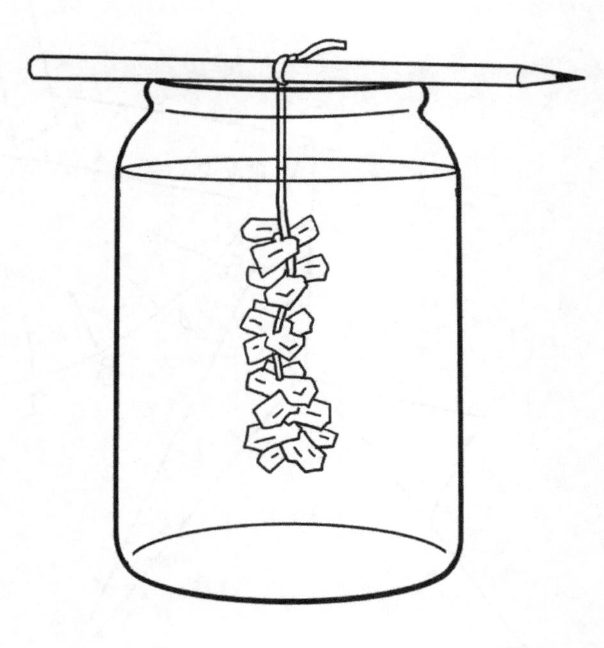

PCM P1: Making a sundial

Copy the sundial template from PCM P2 on to thick card.

Cut along the dotted outlines and fold along the solid black lines.

Stick down the triangular tabs A and B.

Stick on a pointer made from folded paper or card, or insert a dowel rod, at C.

Place your finished sundial on a level surface with the pointer towards north.

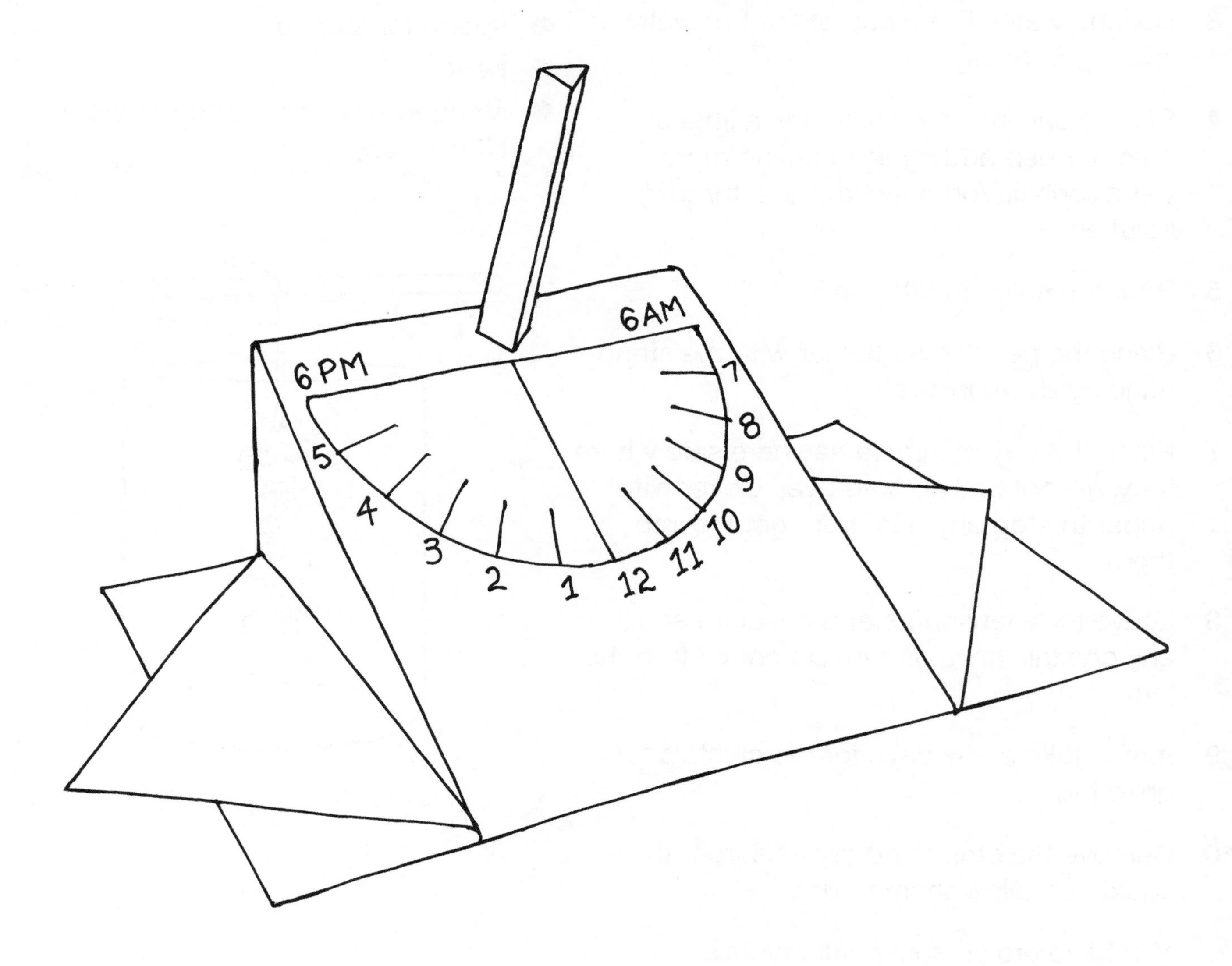

PCM P2: Sundial template

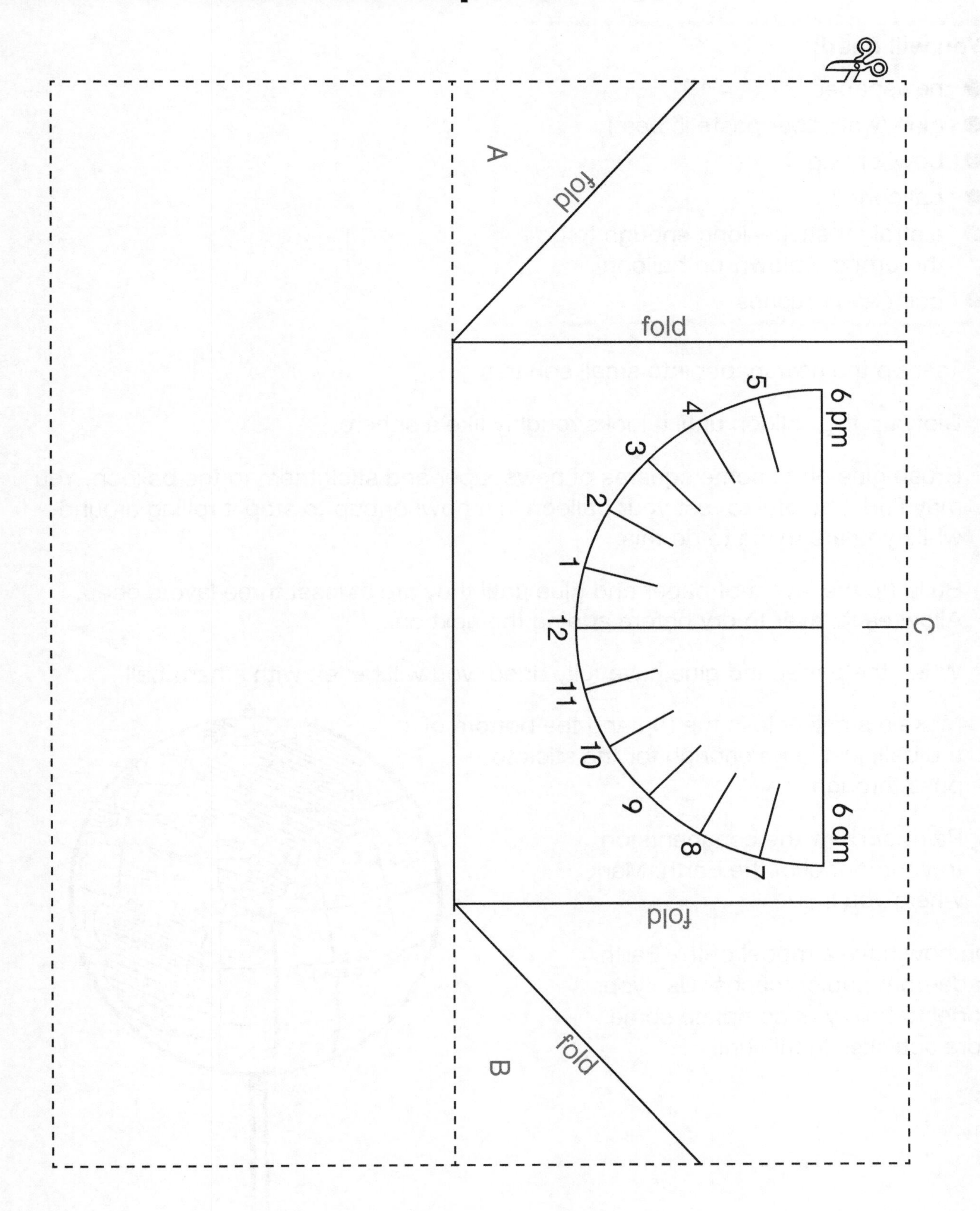

PCM P3: Making a model of the Earth

You will need:

- newspaper
- glue (wallpaper paste is best)
- bowl or cup
- balloon
- a straight stick – long enough to go through the blown-up balloon
- paint and brushes

1 Tear up the newspaper into small squares.

2 Blow up the balloon until it looks roughly like a sphere.

3 Brush glue on to some squares of newspaper and stick them to the balloon. You may find it helpful to rest your balloon in a bowl or cup to stop it rolling around while you are trying to do this.

4 Build up the layers of paper and glue until they are at least three layers deep. Allow each layer to dry before adding the next one.

5 When the paper and glue have fully dried, you will be left with a hard ball.

6 Make a small hole in the top and the bottom of the ball, just large enough for the stick to pass through.

7 Paint or draw the continents on to your model of the Earth. Mark where you live.

You now have a model of the Earth made from papier mâché. Use your model to help you complete some more activities in this unit.

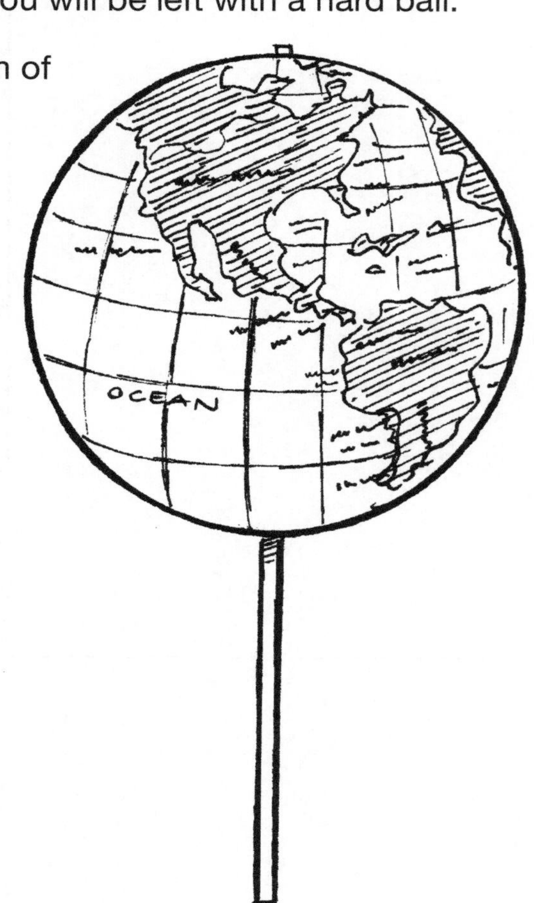

PCM P4: Our spinning planet Earth

Although the Sun seems to move across the sky, it is really the Earth that is moving. You can see how this happens by using yourself as a model Earth.

- Stand facing a friend.
- Hold up one finger in front of your face and look straight at your finger.
- Keep looking at your finger while you turn round slowly.
- Your friend seems to be moving around you, but it is you who are moving!

The Earth spins in the same way. The Sun stays in the same place, but because the Earth is spinning, the Sun seems to be moving, just as your friend seems to move.

Now swap with your partner so that you each have a turn as the Earth.

PCM P5: Mercury planet template

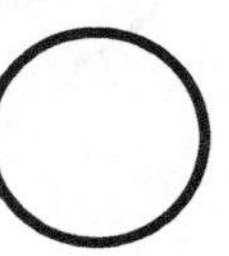

Distance from the Sun: 56.9 million kilometres

PCM P6: Venus planet template

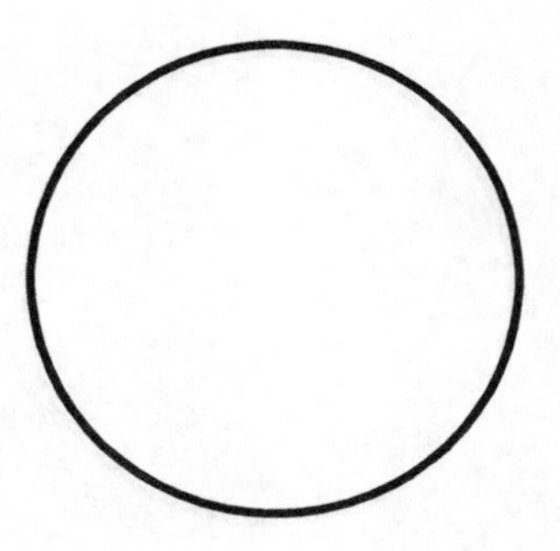

Distance from the Sun: 108 million kilometres

PCM P7: Earth planet template

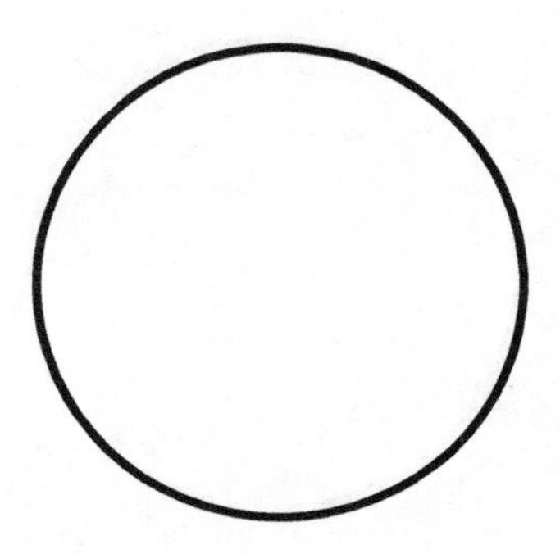

Distance from the Sun: 149.6 million kilometres

PCM P8: Mars planet template

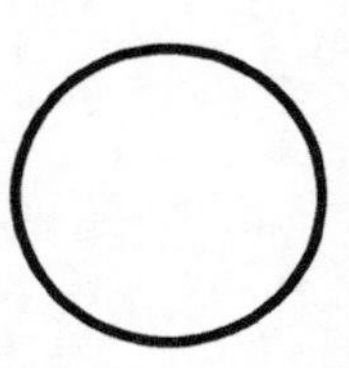

Distance from the Sun: 228 million kilometres

PCM P9: Jupiter planet template

Print this sheet enlarged on A3 paper, or to make it a more accurate representation, draw a large circle with a diameter of 45 cm.

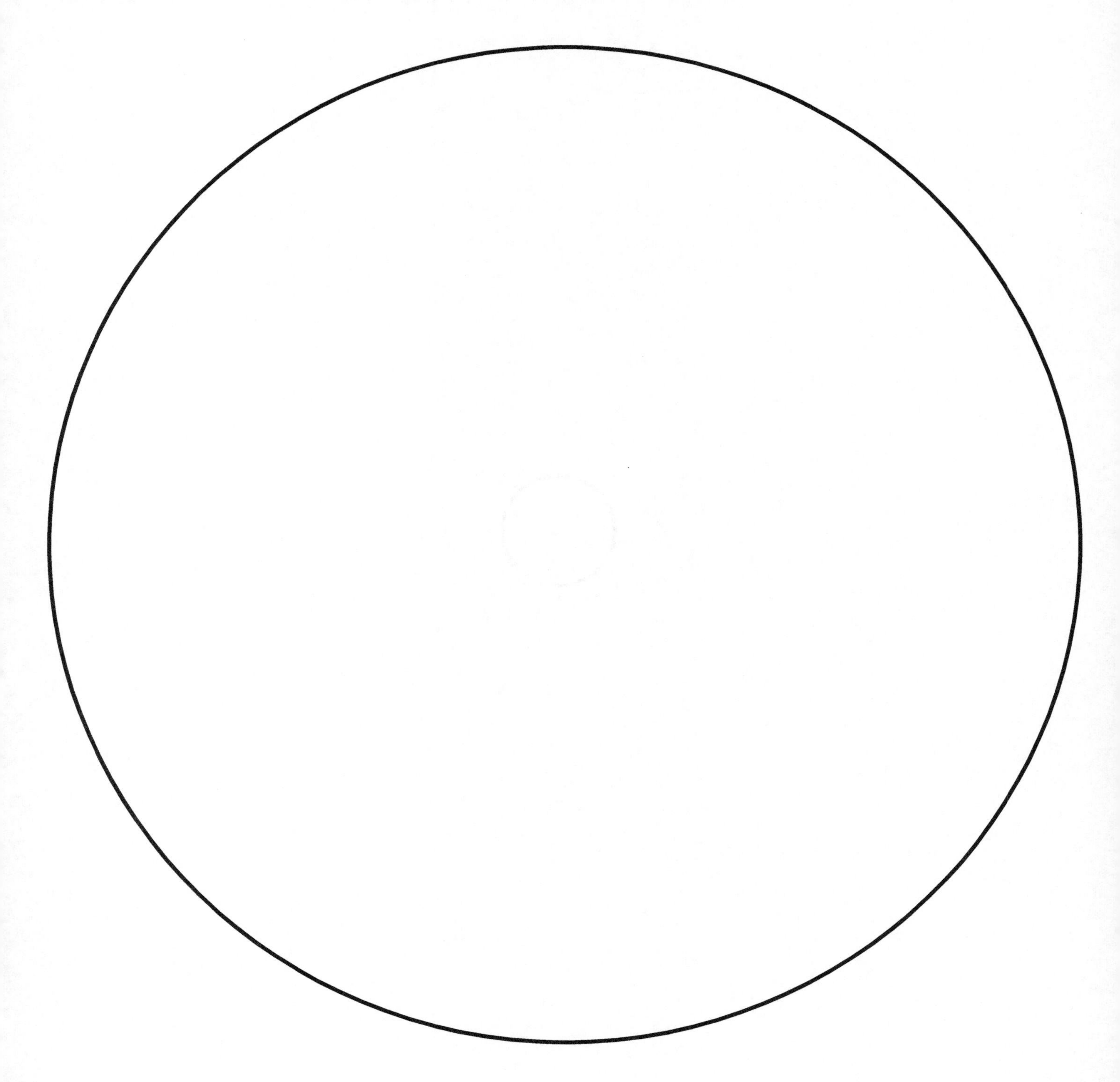

Distance from the Sun: 778.5 million kilometres

PCM P10: Saturn planet template

Print this sheet enlarged on A3 paper, or to make it a more accurate representation, draw a large circle with a diameter of 38 cm.

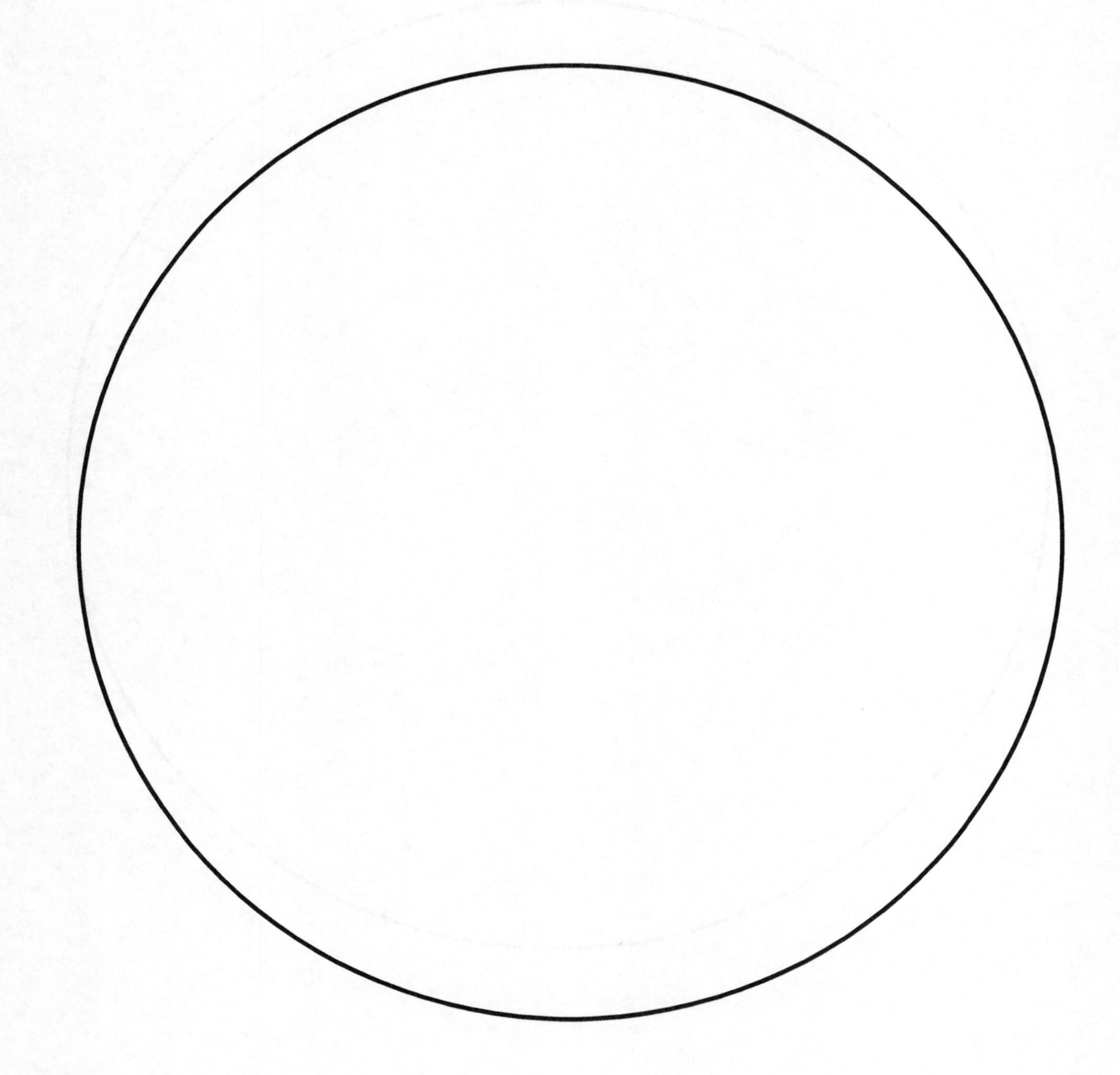

Distance from the Sun: 1427 million kilometres

PCM P11: Uranus planet template

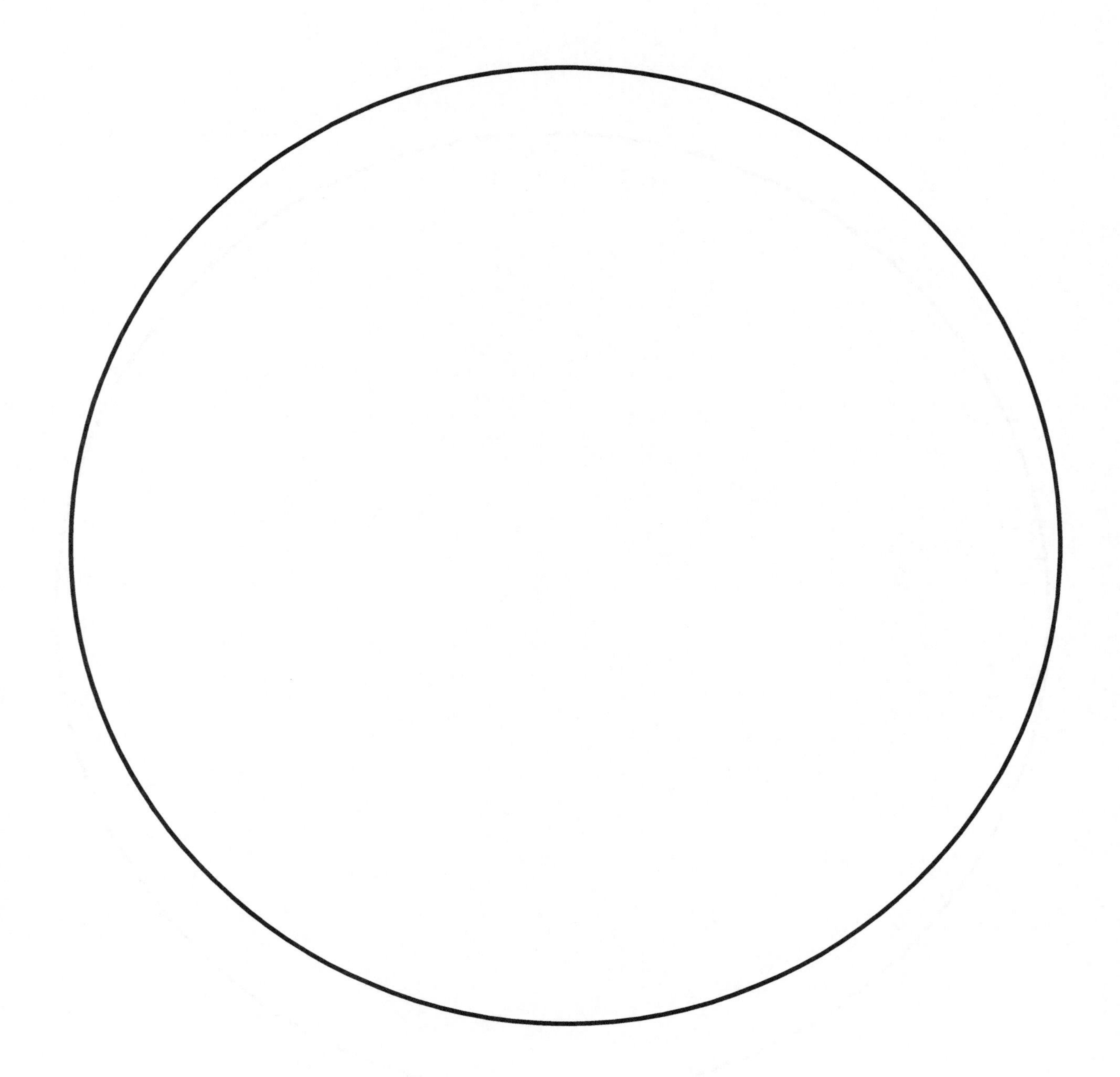

Distance from the Sun: 2870 million kilometres

PCM P12: Neptune planet template

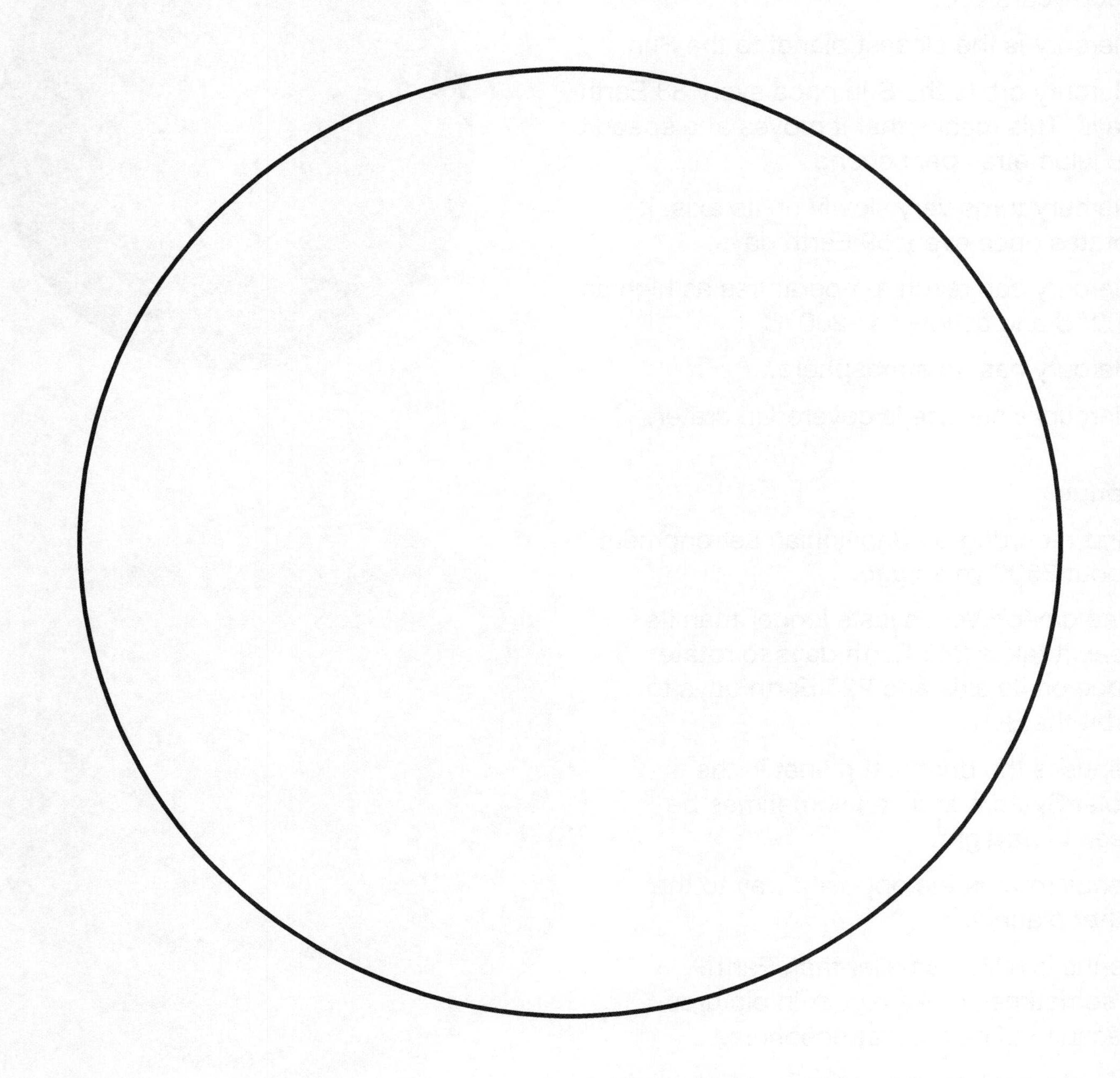

Distance from the Sun: 4497 million kilometres

PCM P13: Mercury and Venus

Mercury

First recorded by Assyrian astronomers about 3000 years ago.

Mercury is the closest planet to the Sun.

Mercury orbits the Sun once every 88 Earth days. This means that it moves at a speed of 48 kilometres per second.

Mercury turns very slowly on its axis. It rotates once every 59 Earth days.

Mercury can reach temperatures as high as 400 °C and as low as –200 °C.

Mercury has no atmosphere.

Mercury's surface is covered in craters.

Venus

First recorded by Babylonian astronomers about 3500 years ago.

One day on Venus lasts longer than its year. It takes 243 Earth days to rotate once on its axis and 225 Earth days to orbit the Sun.

Venus is the brightest planet in the Solar System and can sometimes be seen in daylight.

Venus rotates the opposite way to the other planets.

Venus is a little smaller than Earth. It sometimes looks bigger in pictures because of its thick atmosphere.

The atmosphere on Venus is mainly carbon dioxide. This is a greenhouse gas and so any heat that reaches Venus from the Sun is trapped there.

Venus is the hottest planet and can reach temperatures of 460 °C.

Venus has mountains, craters and volcanoes on its surface.

PCM P14: Earth and Mars

Earth

The Earth takes 24 hours (one day) to rotate on its axis.

The Earth takes 365 days to orbit the Sun.

Earth is the third planet from the Sun.

Three-quarters of the Earth's surface is water.

The Earth measures nearly 13 000 kilometres across.

The Earth has one moon. The Moon orbits the Earth.

The Earth has an atmosphere around it that is mostly nitrogen and oxygen.

The atmosphere helps to keep in the heat that the Earth gets from the Sun.

The Earth supports lots of different animals and plants.

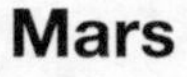

Mars

First recorded by Egyptian astronomers about 3000 years ago.

It takes 687 Earth days to orbit the Sun. One day on Mars is 37 minutes longer than an Earth day.

Mars is sometimes called the 'red planet'.

Mars has the highest mountains and deepest canyons of all the planets. It also has the largest volcano in the Solar System.

Mars has only a very thin atmosphere. This means that it is very cold, and the highest temperature is approximately –5 °C.

Mars has many craters on its surface. These were caused by asteroids and meteorites colliding with the planet.

Mars has polar ice caps which are made mostly of frozen water. There is also some evidence that there may have been oceans and rivers on Mars. This makes scientists think that there may once have been life on Mars.

Mars is being explored by a spacecraft that landed there. It is collecting data and taking samples of rock to bring back to Earth.

PCM P15: Jupiter and Saturn

Jupiter

First recorded by Babylonian astronomers about 2800 years ago.

It takes nearly 12 Earth years to orbit the Sun.

One day on Jupiter lasts about 10 Earth hours.

Jupiter is the biggest planet in the Solar System.

Jupiter is a very stormy planet. There are constant storms, and this makes Jupiter very colourful.

Jupiter has a 'red spot' that is actually a huge storm.

Jupiter is a gas planet. This means it has no solid surface.

If you went to Jupiter, you would weigh two and a half times more than you do on Earth. It also has a very strong magnetic field.

Jupiter has many moons.

Saturn

First recorded by Assyrian astronomers about 2800 years ago.

The rings around Saturn were discovered by Galileo in 1610 AD.

Saturn is the second largest planet.

Saturn has a set of rings around it. These make Saturn easy to recognise in a picture. The rings are made up of particles that may have been left over from a moon that used to orbit Saturn.

It takes Saturn about 10 years to orbit the Sun once.

It takes Saturn approximately ten and a half hours to rotate once on its axis.

Saturn is a gas planet.

Saturn is a very windy planet. That is what gives it the 'swirly' appearance on its surface.

PCM P16: Uranus and Neptune

Uranus

The first planet to be discovered with the use of a telescope.

Discovered by William Herschel in 1781 AD.

It takes 84 Earth years to orbit the Sun.

One day on Uranus lasts about 17 Earth hours.

Uranus is a gas planet.

Uranus is unusual in that it spins on its side.

Uranus has a thick atmosphere of methane, hydrogen and helium.

Much of what we know about Uranus came from a visit by a space probe, Voyager 2, in 1986.

Uranus is sometimes called the 'ice planet' because it is so cold.

There may be a huge ocean on Uranus, but this is not made of just water. This ocean is thought to be very, very hot.

Neptune

Discovered by Urbain Le Verrier and Johann Galle in 1846 AD.

It takes 165 Earth years to orbit the Sun.

One day on Neptune lasts about 18 Earth hours.

Neptune is a gas planet.

Neptune's atmosphere is made of methane, hydrogen and helium.

Much of what we know about Neptune came from a visit by a space probe, Voyager 2, in 1989.

Neptune has the strongest winds of all the planets.

Neptune is extremely cold.

PCM P17: Orbits (1)

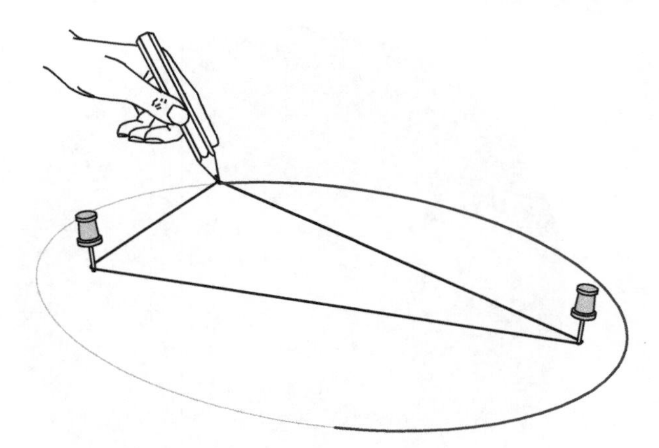

You will need:

- a wooden or cork board
- 2 pushpins
- loops of string of two different lengths
- 3 large sheets of paper
- a pencil

Instructions

1 On the diagram above, add labels to show what represents:
- the Sun
- a comet
- the orbit of the comet.

2 On a large sheet of paper, use the longer loop of string and two pushpins to draw an oval-shaped orbit as shown above. Draw two dots to show where you put the pins. This is your 'control' orbit.

3 On another sheet of paper and still using the long loop, draw another orbit, but with the pins closer together. Draw dots to show where you put the pins this time.

4 Now use the shorter loop of string to draw another orbit on a third sheet of paper. Put the pins the same distance apart as in step 2. Again, mark where you put the pins.

5 Draw sketches of your three orbits in the boxes. (You will need to draw the 'control' orbit twice.) Include dots to show where you put the pins each time.

'Control' orbit	'Pins closer' orbit

PCM P17: Orbits (2)

'Control' orbit	'String shorter' orbit

1 Describe the effect on the shape of the orbit of moving the two pins closer together.

____

____

2 On each of your original diagrams, mark:

- where the comet is closest to the Sun
- where the comet is furthest from the Sun.

Biology: Assessment Sheet B1

1 Read the statements. Circle either 'True' or 'False'.

Seeds need light, water and food to germinate. TRUE / FALSE

Seeds need water and warmth, but not light, in order to germinate. TRUE / FALSE

Seeds are formed when pollen fertilises the ovum of a flower. TRUE / FALSE

[3 marks]

2 Complete the sentences using the words in the box.

fertilisation	insects	anthers	stigma

Pollen is made in the __________ of a flower. Pollen is taken to other plants by __________. The pollen lands on the __________ of a flower . __________ happens when pollen and an ovum join together to form a seed. [4 marks]

3 Explain why it is important that plants reproduce.

____ [1 mark]

4 Read the statements. Circle either 'True' or 'False'.

Pollination occurs when pollen joins with the ova in the ovary to form seeds. TRUE / FALSE

Pollination occurs when pollen from the anther moves to the stigma. TRUE / FALSE

[1 mark]

5 Name these stages in the life cycle of a plant. Use the correct words from the box.

fertilisation	dispersal	pollination	growth

Seeds drop and fall on the ground in different places. __________

Seeds send up small shoots out of the ground. __________ [2 marks]

continued →

Biology: Assessment Sheet B1 (continued)

6 Give an example of a good pollinator. Explain your answer.

____

____ [2 marks]

7 Look at this life cycle diagram of a flowering plant. Write in the missing labels.

Life cycle of a bean

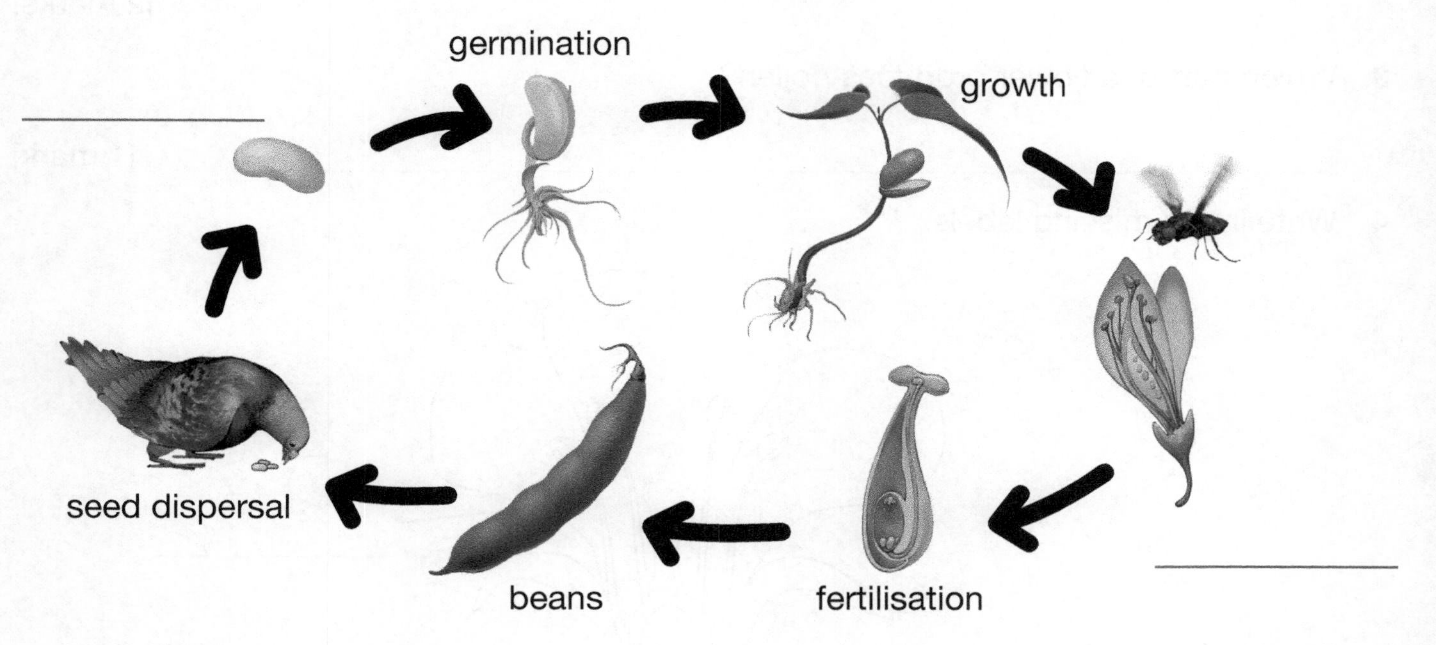

[2 marks]

[Total: ______/15]

Biology: Assessment Sheet B2

1 In which part of a plant do seeds develop? Tick (✓) the correct answer.

the roots ☐ the ovary ☐ the leaves ☐ [1 mark]

2 Read the statements. Circle either 'True' or 'False'.

Plants make their own food using light energy from the Sun. TRUE / FALSE

Pollination occurs when pollen joins with the ovum in the ovary. TRUE / FALSE

[2 marks]

3 Which part of a flower produces pollen?

____ [1 mark]

4 Write in the missing labels.

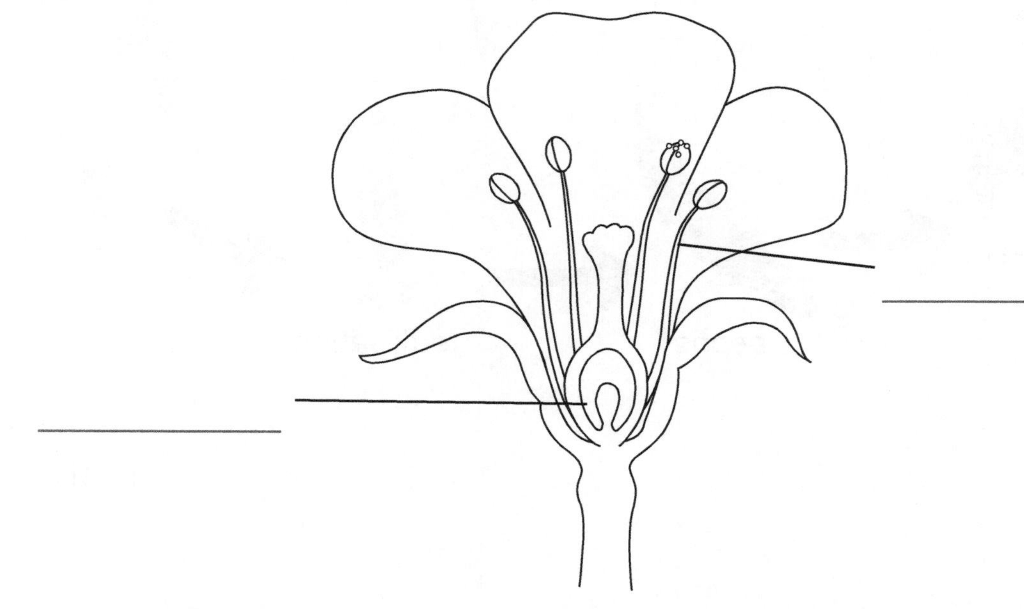

[2 marks]

5 Describe the differences between the growing conditions that seeds need and the growing conditions that plants need.

____

____

____ [4 marks]

continued →

Biology: Assessment Sheet B2 (continued)

6 Circle the correct words.

Seeds can be dispersed by:

a water / weight
b animals / anthers
c words / wind
d exciting / explosion. [4 marks]

7 Circle what a seed does NOT need to germinate.

water light warmth [1 mark]

[Total: ______/15]

Biology: Assessment Sheet B3

1 Fill in the table to sort these parts of a flower into male and female parts.

carpel stamen filament ovary anther ovum stigma style

Female parts	Male parts

[8 marks]

2 Circle the words that are to do with the dispersal of seeds.

wind water rain exploding animals sunlight bees

[4 marks]

3 Draw a diagram to show that the life of a plant continues in a cycle and does not stop.

[3 marks]

[Total: ______ /15]

Biology: Assessment Sheet B4

1 Read the statements. Circle either 'True' or 'False'.

To germinate means to grow flowers and fruit.	TRUE / FALSE
Only a few of the seeds that fall on the ground will germinate.	TRUE / FALSE
Seeds need water to germinate.	TRUE / FALSE
Seeds also need light and warmth to germinate.	TRUE / FALSE
Plants need light to grow.	TRUE / FALSE
Seeds usually germinate under the ground.	TRUE / FALSE

[6 marks]

2 Draw a line from the beginning of each sentence to the end.

Pollen is made…	…attract insects.
The colourful petals of a flower…	…by the anthers.
Bees are good…	…develops into a fruit.
The fertilised part of the flower…	…pollinators.

[4 marks]

3 Circle the correct words.

Pollen / **petal** grains have little **hands** / **hooks** on them, which make them sticky. They stick to the **clothes** / **bodies** of the insects that land inside the flower. Some of the pollen falls off the insect and on to the **ground** / **stigma**, which is also sticky. Sometimes the pollen falls on the **carpel** / **stigma** of the same flower and sometimes it falls on the stigma of flowers nearby as the insects move from flower to flower.

[5 marks]

[Total: ______/15]

Chemistry: Assessment Sheet C1

1 Read the statements. Circle either 'True' or 'False'.

When water condenses it changes from a gas to a liquid. TRUE / FALSE

The boiling point of water is 1000° C. TRUE / FALSE

When a liquid evaporates from a solution it leaves behind a solid. TRUE / FALSE

[3 marks]

2 Complete this paragraph about evaporation. Use the words in the box.

surface	air	gases	temperature

Liquids can change into __________ through the process of evaporation. Factors such as __________, the flow of air and the __________ area can affect the rate of evaporation. Gases go up into the __________ during evaporation. [4 marks]

3 Explain why condensation is the reverse of evaporation.

____

____ [2 marks]

4 Read the statements. Circle either 'True' or 'False'.

When water vapour is cooled down, the water vapour changes into a liquid. TRUE / FALSE

A gas changes into a liquid when the temperature increases. TRUE / FALSE

[1 mark]

5 Complete this sentence.

Water in a solid state melts and turns into a liquid at a temperature of __________°C.

[1 mark]

continued →

Chemistry: Assessment Sheet C1 (continued)

6 Complete the diagram of the water cycle. Add arrows to show how the water moves.

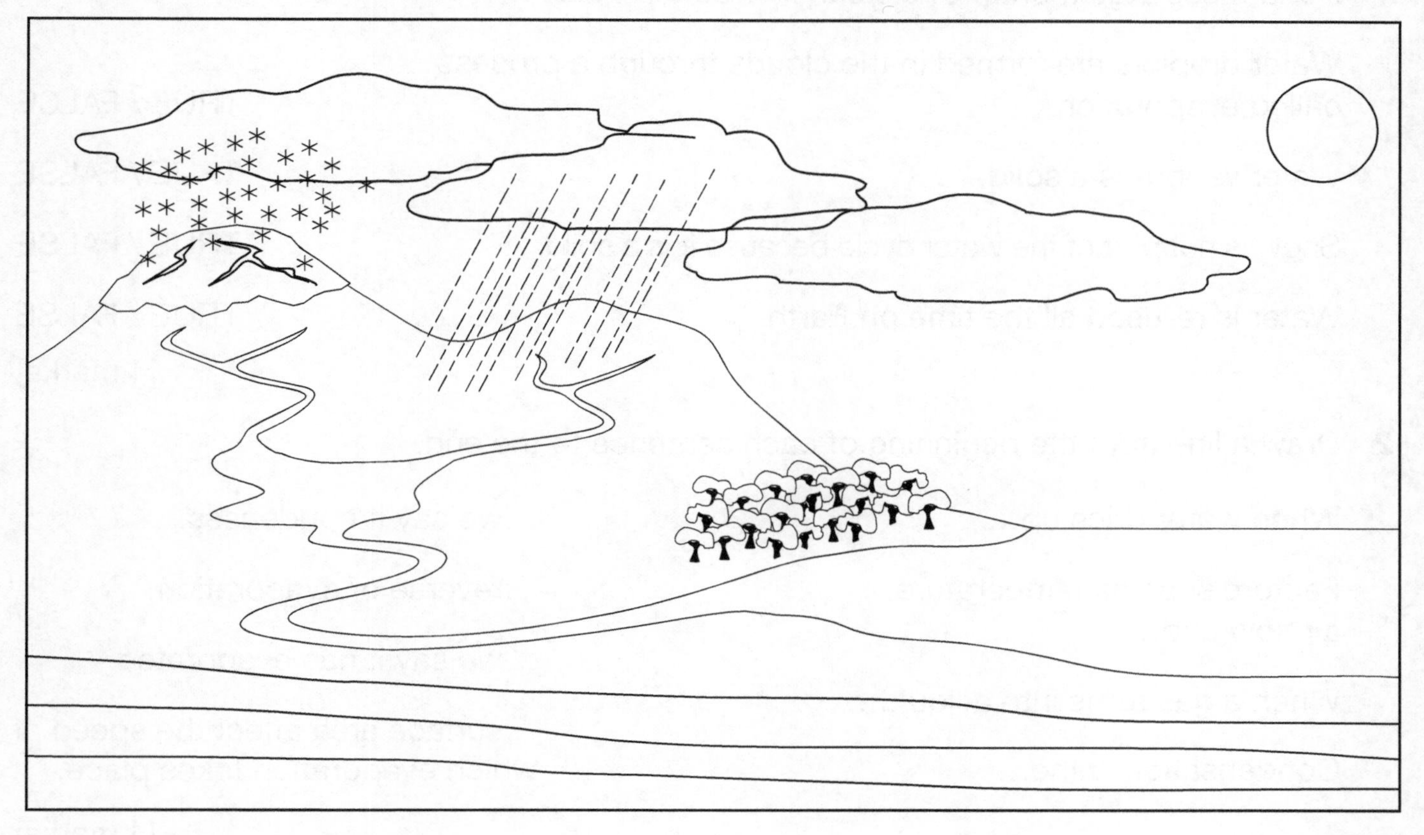

[4 marks]

[Total: ______/15]

Chemistry: Assessment Sheet C2

1 Read these statements. Circle either 'True' or 'False'.

Water droplets are formed in the clouds through a process called evaporation. TRUE / FALSE

Water vapour is a solid. TRUE / FALSE

Snow is not part of the water cycle because it is a solid. TRUE / FALSE

Water is re-used all the time on Earth. TRUE / FALSE

[4 marks]

2 Draw a line from the beginning of each sentence to the end.

When water dries up…	…we say it condenses.
Factors such as temperature, air flow and…	…reverse of evaporation.
When a gas turns into a liquid…	…we say it has evaporated.
Condensation is the…	…surface area affect the speed at which evaporation takes place.

[4 marks]

3 Circle the correct words.

Some substances dissolve in liquids to make **solutes / solutions**.

A substance that can dissolve is called a **solution / solute** and the

liquid in which it dissolves is called the **solvent / solution**.

Water is a common solvent. [3 marks]

continued →

Chemistry: Assessment Sheet C2 (continued)

4 Name three things that can affect the speed of evaporation.

____

____

____ [3 marks]

5 Fill in the missing word.

Most __________ turn to liquids if they are cooled. [1 mark]

[Total: ______/15]

Physics: Assessment Sheet P1

1 Read the statements. Circle either 'True' or 'False'.

Light travels in straight lines. TRUE / FALSE

We can measure the direction of light in lux. TRUE / FALSE

Shadows are always sharp and black. TRUE / FALSE

We can see because light enters our eyes directly from a source or because light is reflected to our eyes. TRUE / FALSE

A beam of light can change direction when it is reflected off a shiny, smooth surface. TRUE / FALSE

[5 marks]

2 Complete this paragraph about shadows. Use the words in the box.

blocked	position	light

Shadows are formed when __________ travelling from a source

is __________. The size of a shadow is affected by the object's

__________ (where it is). [3 marks]

3 Explain the difference between transparent and translucent materials.

____

____ [2 marks]

4 Read the statements. Circle either 'True' or 'False'.

Opaque materials let some light through them. TRUE / FALSE

Opaque materials let no light through them. TRUE / FALSE

[2 marks]

continued 

Physics: Assessment Sheet P1 (continued)

5 Look at this picture and complete the sentence to explain how the person can see over the wall.

There are __________ inside this periscope which __________ the light to our eyes.

[2 marks]

6 Which diagram shows how light is reflected off a shiny, smooth surface, A or B?

A

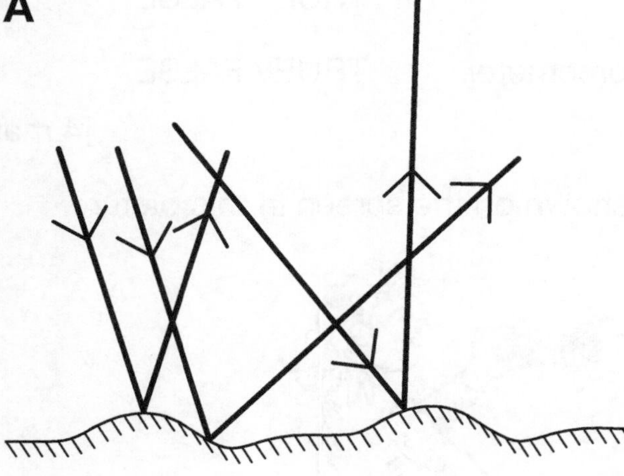

B

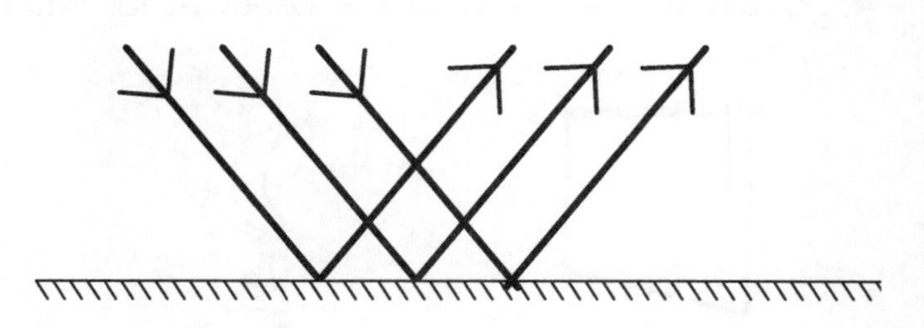

[1 mark]

[Total: ______/15]

Physics: Assessment Sheet P2

1 Name three ways in which shadows outside change during the day.

____

____

____ [3 marks]

2 Circle the correct words.

We can see objects because light from **the Moon / a light source** shines on the objects, is **reflected off / absorbed by** them and then enters our eyes. If light bounces off a surface, we say the light is **absorbed / reflected** by the surface. [3 marks]

3 Read the statements. Circle either 'True' or 'False'.

Shadows are smaller when an object is closer to the light source.	TRUE /FALSE
Shadows are smaller when an object is further away from the light source.	TRUE /FALSE
The lenses in sunglasses are translucent.	TRUE / FALSE
To measure light accurately, we can use a forcemeter.	TRUE / FALSE

[4 marks]

4 Draw the shadow of the object that will be shown on the screen in this picture.

[1 mark]

continued 

Physics: Assessment Sheet P2 (continued)

5 Draw two arrows to show how the light travels from the light source to X in this picture.

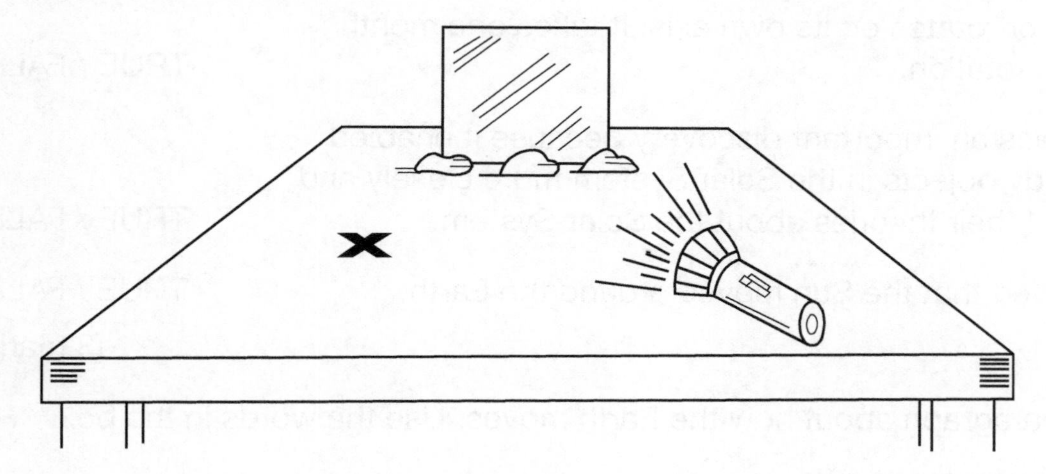

[2 marks]

6 Complete this paragraph about seeing light. Use the words in the box. (You do not need to use all of the words.)

darkness	light	reflected	races

The point where __________ enters the eye is called the pupil.

It looks like a black spot but it is actually a hole. We can see

objects because light from a light source shines on an object,

it is __________ off them and then enters the eye. [2 marks]

[Total: ______/15]

Physics: Assessment Sheet P3

1 Read the statements. Circle either 'True' or 'False'.

The Earth spins or rotates on its own axis. It takes one month to complete one rotation. TRUE / FALSE

The telescope was an important discovery because it enabled scientists to study objects in the Solar System more closely and to prove some of their theories about the Solar System. TRUE / FALSE

Copernicus proved that the Sun moved around the Earth. TRUE / FALSE

[3 marks]

2 Complete this paragraph about how the Earth moves. Use the words in the box.

orbit	axis	hours	days

The Earth spins on its own __________. It takes 24 __________ to make one complete turn or rotation. The Earth also moves in an __________ around the Sun. It takes the Earth about 365 __________ to complete one orbit. [4 marks]

3 Circle the correct words.

The Earth is shaped like a **saucer** / **sphere**. It moves around the **Moon** / **Sun** in **an anti-clockwise** / **a clockwise** direction (from west to east) and it also **jumps** / **spins** as it moves. [4 marks]

continued 

Physics: Assessment Sheet P3 (continued)

4 Explain in your own words why we have daytime and night-time on the Earth.

____

____

____

____ [2 marks]

5 Name one important scientist who made a discovery about the Solar System, and describe their discovery.

____

____

____

____ [2 marks]

[Total: ______/15]

Physics: Assessment Sheet P4

1 Label this diagram of the Solar System. Use the words in the box.

Sun	Earth	Neptune	Mars	Jupiter
Mercury	Saturn	Venus	Uranus	

[9 marks]

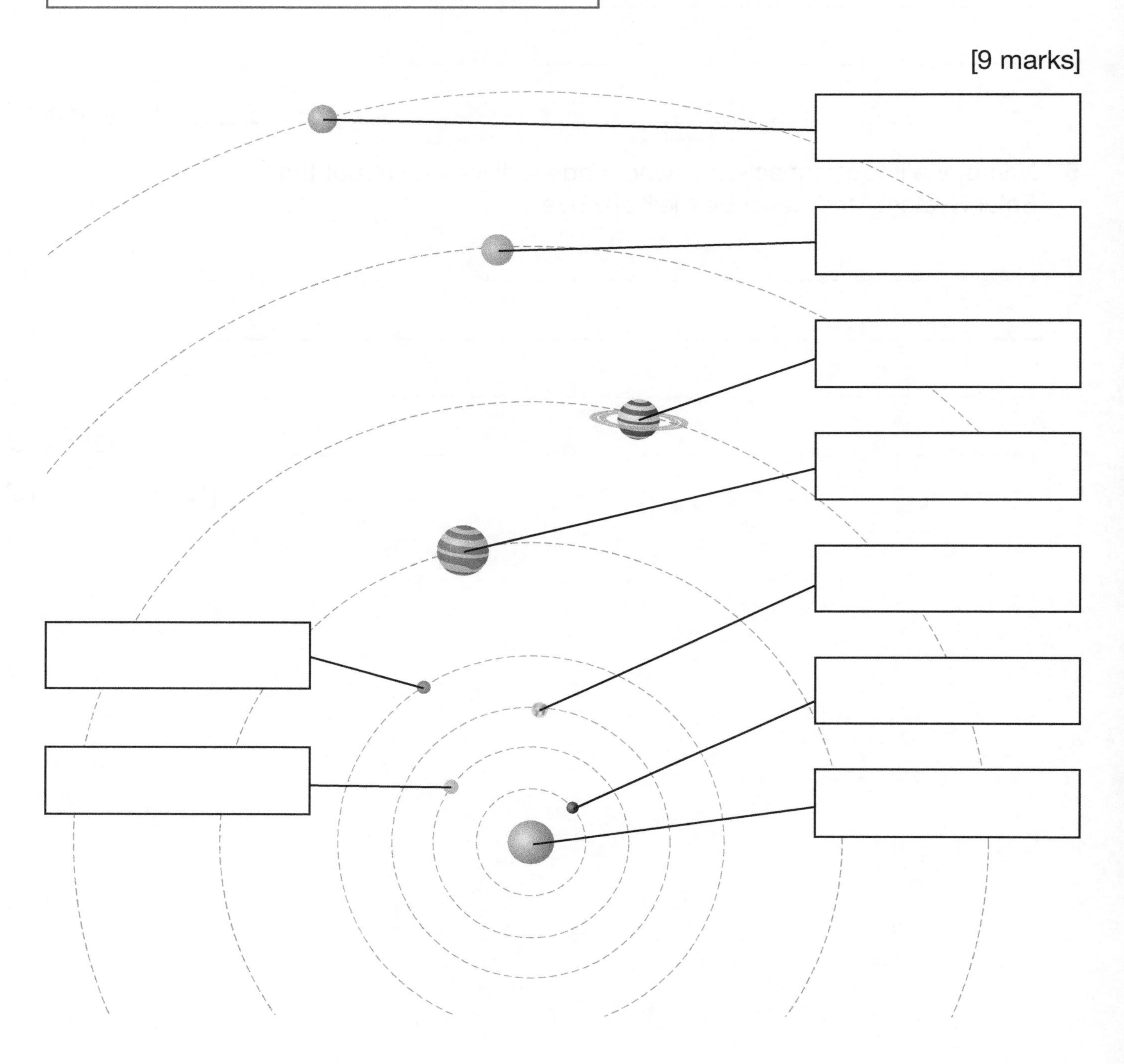

continued

Physics: Assessment Sheet P4 (continued)

2 Look at the diagram and read the sentences under it. Circle either 'True or 'False'.

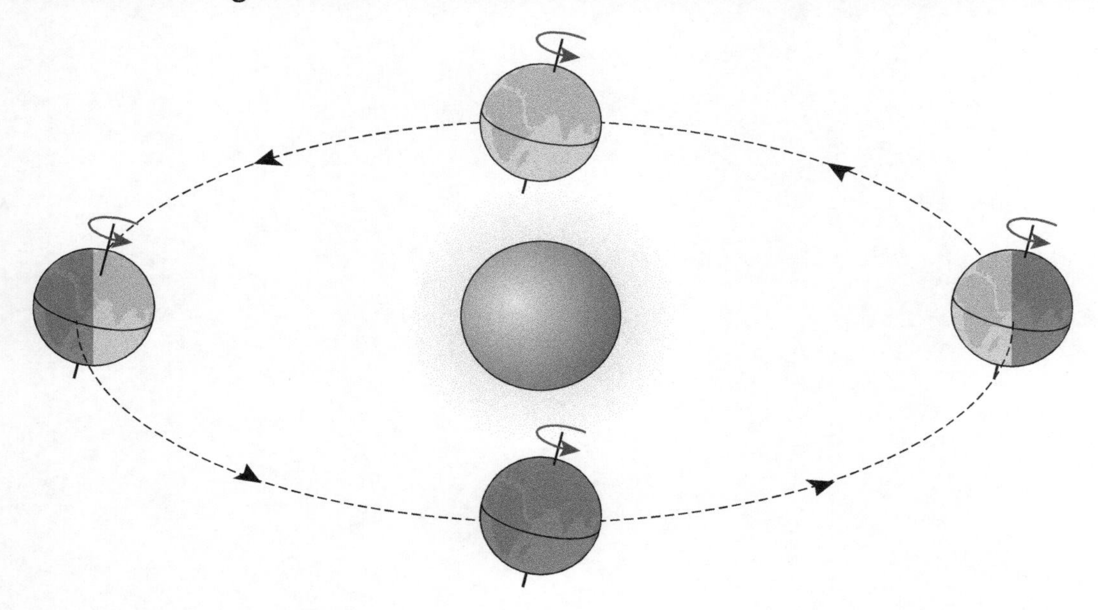

There are four planets that move around the Sun. TRUE / FALSE

The Earth spins on its own axis as it orbits the Sun. TRUE / FALSE

[1 mark]

3 Read the statements. Circle either 'True' or 'False'

The Earth is the only planet in our Solar System which has perfect conditions for life. TRUE / FALSE

There are four planets in our Solar System. TRUE / FALSE

The constellations and stars we can see in the night sky change throughout the year. TRUE / FALSE

[3 marks]

4 Name two things that human-made satellites can be used for.

____

____ [2 marks]

[Total: ______/15]